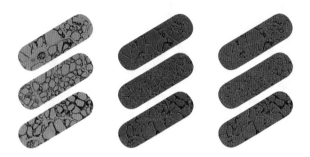

OXFORD BIOLOGY PRIMERS

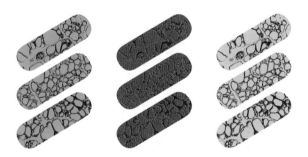

Discover more in the series at

www.oxfordtextbooks.co.uk/obp

Published in partnership with the Royal Society of Biology

THE MARINE ENVIRONMENT AND BIODIVERSITY

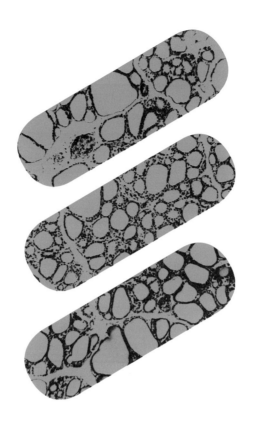

THE MARINE ENVIRONMENT AND BIODIVERSITY

Michael Kent

Edited by Ann Fullick

Editorial board: Ian Harvey, Gill Hickman, Sue Howarth and Hilary Otter

UNIVERSITY PRESS

Great Clarendon Street, Oxford, OX2 6DP,
United Kingdom

Oxford University Press is a department of the University of Oxford.
It furthers the University's objective of excellence in research, scholarship,
and education by publishing worldwide. Oxford is a registered trade mark of
Oxford University Press in the UK and in certain other countries

Published in the United States of America by Oxford University Press
198 Madison Avenue, New York, NY 10016, United States of America

British Library Cataloguing in Publication Data

Data available

Library of Congress Control Number: 2021943511

ISBN 978–0–19–886908–5

Printed in Great Britain by
Bell & Bain Ltd., Glasgow

PREFACE

Welcome to the Oxford Biology Primers

There has never been a more exciting time to be a biologist. Not only do we understand more about the biological world than ever before, but we're using that understanding in ever-more creative and valuable ways.

Our understanding of the way our genes work is being used to explore new ways to treat disease; our understanding of ecosystems is being used to explore more effective ways to protect the diversity of life on Earth; our understanding of plant science is being used to explore more sustainable ways to feed a growing human population.

The repeated use of the word 'explore' here is no accident. The study of biology is, at heart, an exploration. We have written the Oxford Biology Primers to encourage you to explore biology for yourself–to find out more about what scientists at the cutting edge of the subject are researching, and the biological problems they're trying to solve.

Throughout the series, we use a range of features to help you see topics from different perspectives.

Scientific approach panels help you understand a little more about 'how we know what we know'–that is, the research that has been carried out to reveal our current understanding of the science described in the text, and the methods and approaches scientists have used when carrying out that research.

Case studies explore how a particular concept is relevant to our everyday life, or provide an intimate picture of one aspect of the science described.

The bigger picture panels help you think about some of the issues and challenges associated with the topic under discussion–for example, ethical considerations, or wider impacts on society.

More than anything, however, we hope this series will reveal to you, its readers, that biology is awe-inspiring, both in its variety and its intricacy, and will drive you forward to explore the subject further for yourself.

ABOUT THE AUTHOR

Dr Michael Kent MemMBA FRSB studied Zoology at London University, Marine Biology at Bangor, and carried out PhD research into shellfish parasites at Plymouth. After his PhD, he joined Torpoint School as a science teacher, then moved to St Austell VI Form College to teach A-level Biology and Sport Science. From St Austell, he was appointed Head of the Centre for Applied Zoology at Newquay (now part of Cornwall College) where he helped design and deliver Foundation Degrees in Marine Aquaculture and Zoological Conservation. In 2006, he became a full-time science writer and independent researcher. His books include *Advanced Biology* and the *Oxford Dictionary of Sports Science and Medicine,* both published by OUP. His research focuses on intertidal ecology. He has a passion for all things marine, and enjoys sharing that passion with others.

ACKNOWLEDGEMENTS

First, I would like to thank Tom Ireland, editor of *The Biologist* (the magazine of the Royal Society of Biology) for introducing me to Lucy Wells, publishing and commissioning editor for OUP, and Ann Fullick the editor of the Oxford Biology Primer Series. It was on reading a copy of *The Biologist* that I first became aware of the Oxford Biology Primers which, to borrow some of the words of Sir Paul Nurse (past president of the Royal Society), are designed to inspire students '... to want to understand more about the world around them and in so doing set them on the path of exploration and enquiry.'

I would like to express my sincere thanks to Lucy and Ann for enabling me to be part of the team that has produced a Primer on the ocean part of the world. They have been a joy to work with.

Ann has been a tremendous help throughout the publication process, from formulating proposals, to checking artwork, and correcting proofs. In particular, I would like to thank Ann for encouraging me during some difficult times; this book would be much poorer if it were not for her excellent work as Editor and encourager.

I would also like to express my heartfelt thanks to my daughter Kerris for her detailed and constructive comments on initial drafts of the book; it has benefited greatly from her creative contributions.

Finally, I would like to thank my wife Merryn, as ever, for her loving support and helpful suggestions.

Michael Kent

Wadebridge, Cornwall.

2021

CONTENTS

Contents

1 THE MARINE ENVIRONMENT: UNITED AND DIVIDED

Images of Earth from outer space have made us aware that we live on a beautiful blue planet dominated by one gigantic body of seawater. To emphasize its global and unified nature, the whole sea covering the Earth has been called the 'World Ocean'.

The World Ocean interconnects the five major oceans of the world. It covers 361.1 million square kilometres, more than 70 per cent of the surface of the Earth. It contains approximately 1.335 billion cubic kilometres of seawater. And it provides a huge three-dimensional living space for an amazing variety of different organisms from microscopic bacteria to massive whales. Marine life, in varying numbers and types, can be found throughout the water column and along the sea floor from freezing polar seas to warm tropical waters and temperate shores.

In recognition of its importance, the United Nations General Assembly has designated 8 June as World Oceans Day. A United Nations webpage promoting World Oceans Day gave the following answer to the question 'Why celebrate World Oceans Day?'

We celebrate World Oceans Day to remind everyone of the major role the oceans have in everyday life. They are the lungs of our planet, providing most of the oxygen we breathe. The purpose of the Day is to inform the public of the impact of human actions on the ocean, develop a worldwide movement of citizens for the ocean, and mobilize and unite the world's population on a project for the sustainable management of the world's oceans. They are a major source of food and medicines and a critical part of the biosphere. In the end, it is a day to celebrate together the beauty, the wealth and the promise of the ocean.

So, dip your toes in the sea, make contact with the World Ocean, and experience the marine environment and its organisms for yourself (Figure 1.1).

Figure 1.1 On World Oceans Day 2021, ocean science experts focused on the plight of marine ecosystems, such as coral reefs.

© Holger Anlauf

What is the marine environment?

The meaning of the term 'marine environment' depends on the context in which it is used. In this chapter, we are going to use it in its widest sense to refer generally to the abiotic (physical) and biotic (biological) conditions that occur in the global ocean. However, the global ocean does not have just one set of environmental conditions affecting different parts equally. On the contrary, it is so vast that it contains countless different places in which organisms live. Each place has its own set of environmental conditions and can therefore be said to have its own particular marine environment. The environments of adjacent habitats overlap and interact with each other at all spatial scales. Therefore, when discussing the marine environment of a particular habitat, it is essential to define the size of the habitat being considered.

On a regional or large area scale, the global marine environment is commonly divided on the basis of one or more of the following: geography, topography, depth, or the amount of light penetrating. The remainder of this chapter considers the main divisions of the marine environment and some of the organisms that live within them.

Main divisions of the marine environment

Ecologists often divide the terrestrial environment into biomes, such as tundra, temperate grassland, and tropical rainforest, defined mainly in terms of climate and dominant vegetation. Marine biologists and oceanographers have attempted to divide the marine environment into biomes equivalent to those on land, but to date none of these attempts has been adopted by everyone. Nevertheless, climate does have a dramatic effect on the marine environment, enabling it to be divided broadly into polar, temperate, and tropical zones.

Polar zones

The polar zones are characterized by constantly icy cold seas and extreme seasonal fluctuations in the availability of light. You might think that the harsh conditions would make life almost impossible. The winter is truly bleak for marine life. But in the spring and summer these zones support an abundant and diverse marine biota. The alternation of barren winters and fruitful summers brings about some of the most remarkable seasonal migrations on the Earth.

North polar zone

The Arctic Ocean is almost land-locked. In the spring and summer, as rivers and glaciers thaw, freshwater runs into the Ocean. Although ice is solid water and not at all salty, in the winter when sea ice forms it contains brine-filled channels that are in constant contact with seawater. These channels enable the ice to be colonized by a multitude of microorganisms, including phytoplankton that form the base of unique food chains.

Ice algae are not merely trapped passively in sea ice; they are able to actively grow in it. As well as providing food for under-ice zooplankton grazers, ice algae also release dissolved organic matter into the brine channels which is absorbed by bacteria as a source of energy (see Chapter 3).

If the sea ice remains frozen throughout the summer, it provides a stable environment that retains the phytoplankton in the euphotic zone at all times, thus maximizing photosynthesis and primary production. If the sea ice melts, it releases phytoplankton into the sea where they can proliferate and create an algal bloom.

The plankton-rich retreating edge of Arctic sea ice supports a rich community of pelagic organisms including zooplankton, seals, polar bears and marine mammals (Figure 1.2).

South polar zone

The polar zone in the Southern Ocean is regarded as the windiest and most dangerous ocean on the planet. It is bounded by the Antarctic continental landmass to the south and the open seas of the Atlantic, Pacific, and Indian Oceans to the north.

Sea ice forms around the continent each winter. In the summer, this breaks up, drifts north, and melts. Consequently, Antarctic sea ice is usually no more than one year old and only 1–2 metres thick, about half the average thickness of Arctic sea ice. As in the Arctic, algae colonize the sea ice; diatoms can be particularly abundant.

Figure 1.2 An Arctic food web.

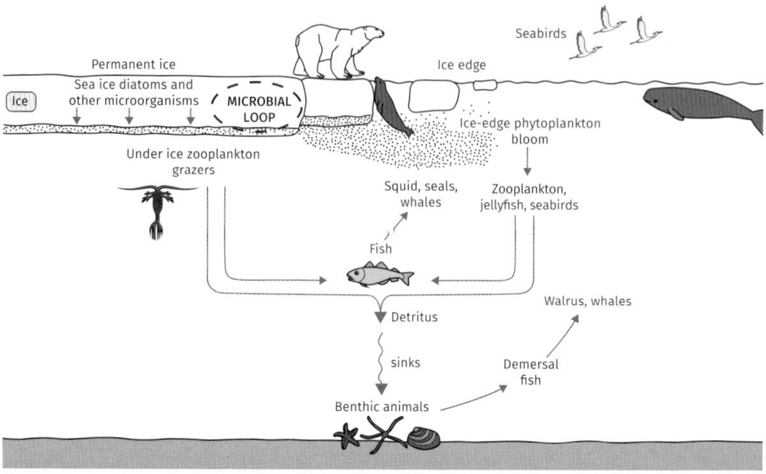

Seabirds

Permanent ice   Ice edge
Sea ice diatoms and
other microorganisms   MICROBIAL
LOOP
Ice   Ice-edge phytoplankton
bloom
Under ice zooplankton
grazers
Squid, seals,   Zooplankton,
whales   jellyfish, seabirds
Fish
Walrus, whales
Detritus
sinks   Demersal
fish
Benthic animals

From: Marine Biology: A very short introduction / P.V. Mladenov, 2013.

Diatoms are among the most exquisite organisms in the marine environment. Each diatom cell is encased within an ornately sculpted, clear glass-like box made of silica. Most diatoms store food as droplets of oil and fatty acids. In some species, individual cells can link to form chains. Diatoms are the dominant phytoplankton in the cold, nutrient-rich waters and sea ice of the Southern Ocean where they form the base of the food webs (Figure 1.3). They occur in sea ice, melt water, and seawater.

Figure 1.3 Light micrograph of a selection of marine diatoms. Each diatom cell is composed of two halves that fit together like a tiny pill box. Their silica-covered surfaces are sculpted with pits, perforations, and striations that form intricate species-specific patterns.

Image: Prof. Gordon T. Taylor/National Oceanic and Atmospheric Administration

The Southern Ocean owes much of its fertility to water movements at the Antarctic Convergence which encircles the northern limits of the Ocean. At the Antarctic Convergence, the very cold, dense surface waters of the Antarctic circumpolar current meet and sink beneath the relatively warm waters of the Atlantic, Pacific, and Indian Oceans. A great mixing of the different waters occurs, resulting in nutrients being brought up to the surface from the sea floor (see Chapter 2). In spring and summer, when the sea ice melts, the release of diatoms into the nutrient-rich sea can lead to massive algal blooms that contribute more than 50 per cent of the primary production in the Southern Ocean (see Case study 1.1, Figure B). A similar phenomenon is seen in other areas, such as the Galapagos.

Case study 1.1
The kingdom of the krill

The southern polar zone has been called the 'kingdom of the krill' because of the dominance of Antarctic krill, an almost transparent shrimp-like creature about 4–6 centimetres in length (see Figure A).

The Antarctic krill (*Euphausia superba*) has a central place in the Southern Ocean food web (Figure B) and is probably the most important species in the Antarctic waters. It feeds on a variety of phytoplankton and zooplankton, but the most important food source is diatoms (see Figure 1.3).

Krill can feed on diatoms that grow on the underside of sea ice and in the open ocean. However, studies using underwater vehicles equipped with cameras have revealed that krill can be more abundant under the ice than in open water, especially during spring. It has been suggested that ice algae are essential for the maintenance of healthy krill populations. If this is correct, any loss of sea ice in the Southern Ocean due to climate change could have significant consequences.

Krill are supremely well adapted for life in the polar zone. They live up to 10 years and cannot migrate any great distance, so they must be able to survive winters in the freezing, dark sea when food is scarce or absent. They do this by lowering their metabolic rate, shrinking their body and assuming a juvenile state by losing reproductive structures. In spring, they are able to grow rapidly and revert to their adult sexually active condition.

Antarctic krill is the main food source for many Southern Ocean predators, including the Blue whale, the largest creature on our planet. It seems remarkable that such a small crustacean could form the staple diet of such a gargantuan creature. But Antarctic krill can form swarms so huge that they are visible from space. It's been estimated that when they are at the peak of their abundance the total mass of Antarctic krill exceeds that of all humans on Earth.

The Blue whale, Right whale, and Fin whale take advantage of this massive source of food by using a sieve-like structure called a baleen in their mouths to capture krill. Dense concentrations of krill, often more than 10 000 per cubic metre, can be scooped up in the open mouth and trapped on the baleen plate. By partly closing the mouth and licking the upper palate with an enormous

Figure A Antarctic krill (*Euphausia superba*).

Photo: pilipenkoD/iStock

Figure B The Southern Ocean food web, dominated by Antarctic krill.

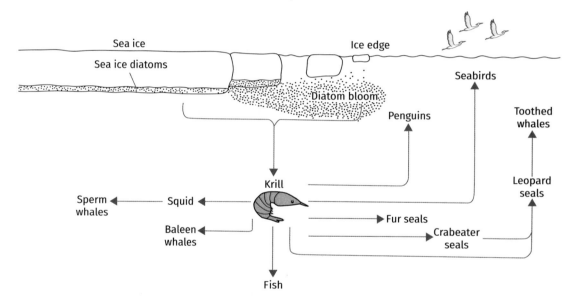

From: Marine Biology: A very short introduction/P.V. Mladenov 2013

tongue, baleen whales force water out of the mouth and swallow the trapped prey. Eaten in bulk with their bodies enriched with oil from diatoms, krill provide a high-energy food for the Blue whale and other predators.

All whales in Antarctica are migratory. They feed in the waters of the Southern Ocean in the summer then go north to breed during the winter months.

❓ Pause for thought

The Southern Ocean food web is more productive and efficient than the Arctic food web. Suggest reasons for this ecologically critical difference.

Krill is a food source for many different predators. Suggest how competition for the same food source might be reduced.

Temperate zones

The temperate zones are between the Tropic of Cancer and the Arctic Circle in the northern hemisphere, and between the Tropic of Capricorn and the Southern Ocean in the southern hemisphere. They are characterized by seas with a temperature range from 10 to 20°C. Seasonal fluctuations in winds and currents drive a strong ocean circulation pattern and an upwelling of the sea that usually result in phytoplankton blooms in spring and autumn. The temperate waters of the North Atlantic support many ecosystems and highly productive fisheries.

Tropical zones

The tropical zones of the marine environment straddle the equator from the Tropic of Cancer in the northern hemisphere to the Tropic of Capricorn in the southern hemisphere. They are characterized by warm surface waters with temperatures rarely falling below 20°C. Marine ecosystems within the tropical zone include coral reefs, mangrove forests, seagrass beds, and vast areas of open ocean.

You might think that warmth, bright sunshine, and a clear blue sea would be a good recipe for the proliferation of marine life. But high productivity occurs in the open ocean of the tropics only where upwelling brings nutrient-rich cool water from the bottom to the surface (see Chapter 2). Where there is no upwelling, the open ocean of the tropics quickly becomes depleted of nutrients. Consequently, biodiversity and productivity are low. Paradoxically, it is in some of these tropical areas depleted of nutrients where coral reefs, renowned for their high biodiversity, occur.

A closer look at coral reefs

Healthy coral reefs are the epitome of a tropical marine paradise. Their beauty, high biodiversity, and productivity make them globally significant. Home to an estimated quarter of all marine species, tropical coral reefs have been called the 'rainforests of the ocean'. As well as supporting marine wildlife, they provide a source of income for millions of people in the tropics. However, coral reefs occupy less than 1.2 per cent of the marine environment; their combined surface area is about the same size as Italy.

Coral reefs grow between 30°N and 30°S, but the communities of those towards the northern and southern extremities tend to be a mixture of coral and seaweeds. It is only in tropical waters that coral reefs can outcompete seaweeds and become the dominant feature.

Coral reefs are dynamic living structures made by colonial animals related to sea anemones (Figure 1.4). Like anemones, coral have tentacles armed with stinging cells which they use to capture prey. Each colony consists of thousands of individuals called polyps that are interconnected and share a common gut cavity. The polyps can reproduce both asexually and sexually, and they build up the reef by extracting calcium from the surrounding seawater to secrete a calcium carbonate skeleton.

Figure 1.4 Anatomy of hard coral. The soft part of the coral has two cellular layers: an outer ectoderm and an inner endoderm separated by a non-cellular jelly-like layer called the mesoglea. The ectoderm of the tentacles contains nematocysts (stinging cells); the endoderm contains zooxanthellae, symbiotic unicellular algae.

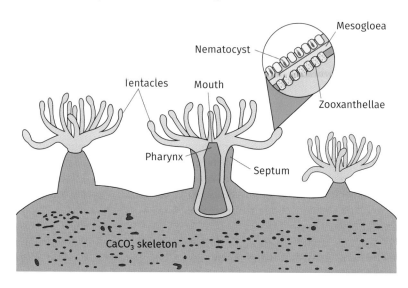

Image: Designua/Shutterstock

Probably the most remarkable feature of reef-building corals is the intimate relationship they form with photosynthetic unicellular algae called zooxanthellae. Reef corals pack zooxanthellae into their tentacles and the lining of the gut, and here the algae are protected and supplied with a constant stream of metabolic waste products—nitrogenous and phosphorous compounds as well as carbon dioxide. If the zooxanthellae are exposed to the right wavelengths of light, they turn these waste products into organic compounds, some of which the coral uses as food. Some coral species obtain as much as 95 per cent of their food from zooxanthellae. Because of this intimate relationship with their photosynthesizing algae, reef-building corals thrive best in warm, clear, and shallow waters. It is the algae which give corals the amazing colours which attract so many tourists.

When corals become stressed, the polyps expel their algal lodgers and the colony takes on a stark white appearance, a phenomenon commonly called coral bleaching. If prolonged, bleaching can lead to the death of the coral, a topic to which we will return later in the book (see Chapter 7).

The pelagic and benthic zones

So far we have looked at climate-based divisions of the world oceans. However, probably the most common way of dividing the marine environment is into the pelagic zone and the benthic zone (see Figure 1.5).

Figure 1.5 The pelagic and benthic zones of the marine environment.

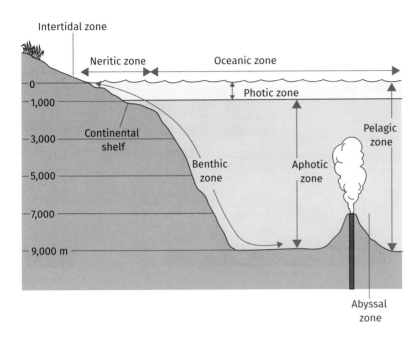

The pelagic zone

The pelagic zone includes the whole body of seawater in the global ocean. It has two main subdivisions: the neritic zone and the oceanic zone. The neritic zone includes intertidal and coastal waters and extends from the top of the shore at high tide to the edge of the continental shelf. It is characterized by having relatively shallow water.

The oceanic zone includes all the deeper water beyond the continental edge. Abiotic conditions change markedly with the depth of water. Whereas pressure increases with depth, light availability decreases until, below a certain depth, no light penetrates at all. Organisms that live in the pelagic zone can generally be placed into two broad categories: nekton and plankton.

Nekton

Nektonic organisms are powerful swimmers roaming more or less freely in the ocean, independent of tides and currents. Typically, the fastest and most powerful, be it a squid, fish, or marine mammal, has a torpedo-shaped body that can cut through the water efficiently. Even penguins and other birds assume this streamlined shape when swimming underwater in search of prey. This similarity in body shape is a result of convergent evolution, driven by the similar lifestyle the members of the nekton share.

Bottlenose dolphins are a good example of a necktonic organism. They are marine mammals with many structural adaptations that enable them to be among the most efficient swimmers in the ocean, and they live a wide-ranging life at sea. They routinely swim at speeds exceeding 10 kilometres

Figure 1.6 Bottlenose dolphin (*Tursiops truncatus*) surfing in the wake of a boat.

Photo: NASA

per hour, and can keep up with boats travelling at almost 30 kilometres per hour.

Although dolphins have hair when in the womb, they lose all of this at birth, or shortly afterwards. A hairless body surface, combined with a torpedo shape, enables dolphins to move through the sea with minimal drag. Like us, and all other mammals, dolphins are air-breathers. One of the major adaptations to a dolphin's life at sea is the modification of nostrils into a single blowhole on the top surface of its head (see Figure 1.6). It is through the blowhole that respiratory gases enter and leave the body. To prevent water flooding the lungs when the dolphin is moving through the sea, the blowhole is covered by a muscular flap which, when relaxed, provides a watertight seal. A dolphin holds its breath while under water and begins to exhale just before reaching the surface by contracting the muscular flap and opening the blowhole. During each breath, a dolphin exchanges 80 per cent or more of its lung air. In comparison, we exchange about 17 per cent with each breath. Reduction of limb bones and the evolution of a dorsal fin and tail fin are just some of the other adaptations that make dolphins such efficient swimmers.

Plankton

Plankton are organisms with limited powers of locomotion—they drift with the currents and tides. They are commonly classified according to their size and whether they photosynthesize (phytoplankton) or not (zooplankton). Phytoplankton have been called the 'grasses of the sea' because they are the main primary producers in the marine environment (see Chapter 6). We often think of plankton as microscopic, but they range in size from viruses and bacteria to jellyfish, as you can see in Table 1.1.

Table 1.1 Size classes of plankton.

Name	Size range	Examples
Femtoplankton	0.02 to 0.2 µm	viruses, small bacteria
Picoplankton	0.2 to 2.0 µm	bacteria
Nanoplankton	2.0 to 20 µm	dinoflagellates
Microplankton	20 to 200 µm	microalgae, juvenile crustacea
Mesoplankton	0.2 to 20 mm	fish larvae
Macroplankton	2 to 20 cm	jellyfish
Megaplankton	>20 cm	large jelly fish

Until the late 1980s, it was generally sufficient to classify phytoplankton and zooplankton simply on the basis of their visibility to the naked eye. Plankton less than 200 µm can only be seen with a good microscope and were grouped together as microplankton. Those above 200 µm were grouped together as macroplankton. This changed after a group of marine scientists revealed the importance of very small planktonic organisms, especially viruses, in the plankton (see Scientific approach 1.1).

Scientific approach 1.1
Revealing marine viruses

In 1989, a team of marine scientists led by Gunnar Bratbak at the University of Bergen used a **transmission electron microscope** (TEM) to count the number of planktonic organisms smaller than 0.2 micrometres in seawater. They found that in spring, samples of seawater from the Norwegian fjords, North Atlantic, and the Barents Sea contained enormous numbers of viruses and bacteria. The results for bacteria were no great surprise, but until then viruses were not thought to be important components of the marine environment; many marine science textbooks barely gave them a mention. But the work of Bratbak and his colleagues revealed that there were 10 million times more viruses in seawater than expected from previous records.

Assuming concentrations of roughly 10 million viruses per millilitre of seawater, it has been estimated that there are about 10^{30} viruses in the global ocean, containing about 200 million tonnes of carbon; this is equivalent to the carbon in about 75 million Blue whales. Whether viruses are living or non-living entities is disputed by some (see Chapter 4)—nevertheless they are included as members of the plankton because of the effect such large numbers can have on other organisms within the marine environment. They have a particularly important role in microbial ecosystems. Before 1989,

Figure A Viruses in a drop of water made visible by epifluorescence microscopy.

Image: Rachel Parsons

the accepted view was that bacterial numbers were controlled from above by protozoans and other organisms grazing on the bacteria. The results of Bratbak and colleagues suggested that bacterial control might come mainly from **bacteriophages**, viruses that attack bacteria. They calculated that as much as a third of the global marine bacterial population might come under attack from phages every day.

❓ Pause for thought

Viruses are regarded by some marine scientists as the most important component of the marine environment, because viral infections can lead to the death of algae, swarms of crustaceans, fish, and whales, as well as bacteria. Suggest how viruses might affect bacteria other than by killing them. How might viruses affect geochemical cycles in the marine environment?

Although the TEM has been superseded by other techniques for counting viruses, such as **epifluorescence microscopy** (Figure A), it continues to be used as it reveals not only the number of viruses in a sample, but also their morphology. This enables the viruses in a sample to be both quantified and identified. However, before using the TEM to count and identify viruses, a suitable sample must be collected from seawater and prepared for viewing under a TEM. How might this be done?

The benthic zone

The benthic zone refers to the whole of the ocean floor. It extends from the intertidal zone (the area between the highest and lowest tides) to the hadal zone at the bottom of the very deepest ocean trenches. Many benthic organisms on rocky shores, such a barnacles and seaweeds, are permanently attached to the substrate. Shore animals and seaweeds are adapted to cope with the extremes of salinity, temperature, and exposure to air that characterize the intertidal zone—there is more on these amazing organisms in Chapter 2 and Chapter 3.

The sublittoral zone extends from the bottom of the intertidal zone to the edge of the continental shelf. It is generally rich in marine life and includes areas with luxuriant growths of seaweeds such as kelp.

The bathyal zone is the portion of the marine environment that extends down the continental slope from a depth of about 200 metres to 4000 metres. Seaweeds are mainly absent, because little light penetrates here for photosynthesis, but fixed animals such as crinoids, sponges, and cold-water coral are often found on rocky outcrops.

The abyssal benthic zone (or abyssobenthic zone) consists mainly of extremely flat and muddy plains that, like the hadal zone, were once thought to be devoid of marine life. The abyssal plains form almost two thirds of the sea floor.

Scientific approach 1.2
The azoic hypothesis and changing mind-sets

This is a deep-sea tale of how the scientific community can become attached to a hypothesis and be very resistant to rejecting it, even when clear contradictory evidence is presented.

We now know that parts of the hadal zone are teeming with life. But in the early nineteenth century, scientists generally believed that no living organism could exist in the deep ocean: it was too dark, too cold, and the pressure was too great. In 1841, British marine scientist Edward Forbes dredged the ocean floor in the Mediterranean and Aegean seas. He found that the number and type of marine species declined progressively with depth of water. Extrapolating from his data, he proposed the azoic hypothesis, stating that life could not exist at depths greater than 600 metres.

Despite reports from ship captains that brittle-stars, benthic worms, and other marine organisms had been brought to the surface on fishing gear and on sounding lines from depths as great as 2 kilometres, the azoic hypothesis was accepted for 25 years. It was not until 1869, when underwater telegraph cables covered with marine fouling organisms were brought to the surface, and scientists on *HMS Porcupine* dredged up deep-sea benthic organisms, that the hypothesis was seriously questioned. The azoic hypothesis was not fully rejected until the historic expedition of *HMS Challenger* (1872–76).

This was the first ship to carry out a comprehensive exploration of the deeper parts of the ocean, as she set off carrying not only scientists but also purpose-built laboratories and equipment for measuring depth and retrieving organisms from the deep ocean (see Figure A).

The *Challenger* expedition discovered marine life at depths close to 6000 metres—and finally debunked the azoic hypothesis for good. The *Challenger* Deep, the deepest part of the World Ocean, is named after this ground-breaking scientific expedition.

Figure A Sailors examining a haul on board the *Challenger*.

Image: W. H. Overend, courtesy of The British Library

? Pause for thought

It is often mistakenly thought that science is about building a body of hard 'facts' that are indisputably and absolutely true. In reality, it's more like writing a special type of story that explains things about the material universe on the basis of the scientific evidence available at the time. One of the great strengths of science is that the truth of the 'story' can be tested by new observations and experiments, and is modified or discarded when sufficient new evidence demands it. Scientific explanations are always open to change. Suggest why Forbes and many other scientists in the nineteenth century were reluctant to reject the azoic hypothesis. Think of another scientific hypothesis that resisted rejection despite evidence against it.

The sea floor

The seafloor is a global feature of the benthic zone. Sailors, fishermen, and divers have long been aware of the topography of the seafloor in shallow waters. Their lives and livelihoods depend on knowing the submarine terrain.

With an average depth of more than 10 000 metres, most of the world's seafloor is beyond the reach of sunlight and is subjected to extremes of cold and pressure. This makes it difficult to explore. Nevertheless, with advances in sonar devices and other investigative tools, we know that the deep sea floor is not as flat and devoid of features as was once thought. On the contrary, it has geological features that are probably more varied and impressive than any found on land; its mountains are the highest and its rift valleys the deepest on Earth. Each ocean is a seawater-filled basin that shares characteristic topographical and geological features (see Figure 1.7).

Figure 1.7 Major geological features of the seafloor.

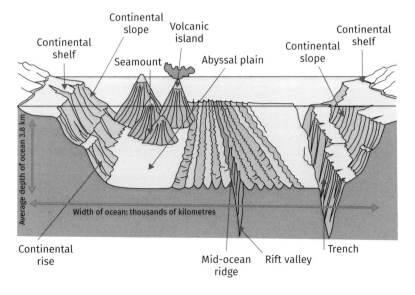

The most prominent features of the sea floor are:

- a gently sloping continental shelf, generally less than 130 metres deep, and from several kilometres to hundreds of kilometres wide;
- continental slope, which drops abruptly to depths of about 3–5 kilometres;
- abyssal plains, vast expanses of flat and soft substrate extending over depths of about 3–5 kilometres; globally, abyssal plains make up about two thirds of the seafloor;
- mid-ocean ridges, underwater mountains that can rise thousands of metres above the surrounding abyssal plains which they transect, and are linked together to form a continuous chain of mountain (65 000 kilometres long) which span all the ocean basins; mid-ocean ridges are created by volcanic activity and are the sites of active hydrothermal vents;
- ocean trenches or canyons which plunge 3–4 kilometres below the surrounding seafloor; they are typically about 1.5 kilometres wide and thousands of kilometres long—much deeper and longer than the Grand Canyon; the Mariana trench (which includes the Challenger Deep) is the deepest part of the World Ocean at nearly 11 kilometres below sea level;
- seamounts, typically extinct volcanoes that generally rise steeply about 1 kilometre above the surrounding seafloor but do not reach the ocean surface; their peaks can be thousands of metres beneath the ocean surface (see Figure 1.8).

Figure 1.8 A diagram of a typical seamount showing the strong currents deflected up its sides. The currents develop into complex spiral eddies that create so-called 'Taylor cones' which carry nutrients up from deep water. The nutrients support marine plankton and local pelagic fish that attract a variety of fishing boats.

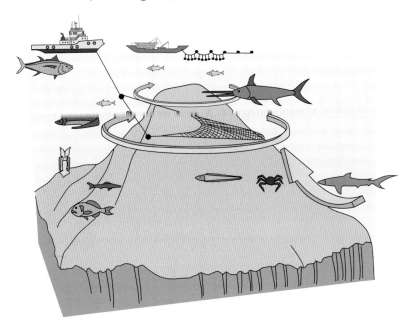

Image: Captain Les Gallagher

A closer look at a seamount: a marine biodiversity 'hotspot'

The Dellwood Seamount is an isolated island-like submarine feature that rises from a depth of 2359 metres in the North East Pacific. Its summit is between 300 and 1218 metres below the ocean surface. Dellwood is unusual for a seamount in Canadian waters in that it has hydrothermal vents, albeit inactive ones. There are estimated to be over 100 000 seamounts in the World Ocean, half of them in the Pacific.

Like many other seamounts, Dellwood has such an extraordinary variety of species that it is regarded as a special kind of deep-sea biological hotspot. Ocean currents that flow unimpeded along the adjacent abyssal plain are deflected up the flanks of the seamount resulting in a localized upwelling. This brings oxygen and nutrients from the ocean depths to the photic zone, enabling phytoplankton to proliferate. Interactions between the upwelling and horizontal currents can result in complex rotating currents that retain and concentrate nutrients and phytoplankton above the seamount.

The phytoplankton form the base of a food web above the seamount that sustains an abundance of zooplankton. In turn, the zooplankton sustain huge shoals of fish and draw in other predators including sharks and seabirds.

The summit and flanks of the seamount are covered with a dense community of suspension feeders that includes cold-water corals, sea fans, brittle starfish, and sponges. Together they create a thicket-like habitat for a host of other species including rockfish and large shoals of halibut.

The bigger picture 1.1
Hydrothermal vents—challenging paradigms

In his book *The Structure of Scientific Revolutions*, Thomas Kuhn defined scientific paradigms as 'universally recognized scientific achievements that, for a time, provide model problems and solutions for a community of practitioners, i.e., what is to be observed and scrutinized'. Put more simply, a scientific paradigm is a widely accepted belief or concept such as 'all major ecosystems on Earth are ultimately dependent on sunlight as a source of energy'. The discovery of hydrothermal vents in 1977 led to this and several other biological paradigms being challenged (see Figure A).

When scientists on the submersible 'Alvin' discovered the hydrothermal vents at a depth of about 2700 metres off the Galapagos Islands, they were totally amazed to find dense populations of worms, bivalve molluscs, crabs, and even fish living close to black smokers. A rich community of organisms living in a marine environment of scalding hot water, toxic chemicals, and no sunlight was totally unexpected. According to all the textbooks at the time, this could not happen. The fact that it does is down to archaea and bacteria forming the base of a food web that functions in total darkness where there is no photosynthesis. Most archaea and bacteria thrive around hydrothermal

Figure A A venting 'black smoker', so-named because it emits jets of particle-laden fluids. The particles are mainly fine-grained sulfides. On cooling, the fluids solidify to form chimney-like iron sulfide structures.

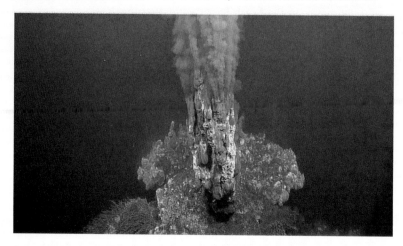

Photo: Ocean Exploration Trust/Ocean Networks Canada

vents because they can use the abundant hydrogen sulfide contained within the surrounding fluids as a source of energy. Using oxygen in the seawater, they oxidize the hydrogen sulfide to make energy available to power the production of organic matter from carbon dioxide and water. This **aerobic chemosynthesis** is similar to photosynthesis in having many steps, but the overall process can be summarized as

$$H_2S + CO_2 + O_2 + H_2O \rightarrow CH_2O + H_2SO_4$$

Although this equation is a gross simplification, it does show the main reactants and products involved.

Some chemosynthetic microorganisms associated with vents can use methane as a carbon source. How they do this is not entirely clear, but as no oxygen is involved in the process it is called anaerobic chemosynthesis. Anaerobic chemosynthesis has enabled ecosystems with metazoan-containing communities to form around vents in the absence of oxygen.

Chemosynthesis is not confined to hydrothermal vents; it occurs in the stomachs of mammals. But the extraordinary feature of hydrothermal vent chemosynthesis is that it supports a whole ecosystem. Some vent animals feed directly on the archaea and digest them in their guts. Others depend on an intimate symbiotic relationship to obtain nourishment. For example, *Riftia pachyptila*, a giant gutless polychaete worm, absorbs hydrogen sulfide and oxygen through its tentacles and into its bloodstream which delivers these raw materials of chemosynthesis to its symbiotic partners. The bacteria manufacture organic matter, some of which is absorbed by the worms.

Up until the discovery of hydrothermal vent ecosystems, it was generally accepted that life arose from simple molecules that formed a broth in the primordial seas that covered much of the Earth. The theory received support from Stanley Miller and Harold Urey's 'lollipop' experiments. These showed that organic molecules could be obtained by the action of simulated lightning on a mixture of the gases containing methane, ammonia, and hydrogen which were thought at that time to represent Earth's earliest atmosphere.

The discovery of hydrothermal vent ecosystems has changed how we view the history of life on Earth. Globally, underwater vents are abundant and are important sources of many elements and organic compounds that are transferred into the surrounding sea. We have seen how they form rich present-day ecosystems without the need of light for photosynthesis. But active hydrothermal vents probably existed as soon as liquid water accumulated on the Earth more than 4.2 billion years ago. It is possible that present-day hydrothermal vent communities include **relict organisms** that resemble the earliest organisms on Earth. Life may not have evolved in this way, but there is a growing group of scientists who believe it did. One such group is a University College London (UCL) research team. In 2019, they created **protocells** in alkaline seawater, supporting the hypothesis that life originated in deep-sea hydrothermal vents. A protocell is a self-assembling spherical collection of lipids. It is seen as a key stepping stone to the development of cell-based life.

❓ Pause for thought

How might studies such as those carried out by the UCL research team help in the search for extraterrestrial life?

Ocean light zones

Light from the sun is the primary source of energy for all life on Earth. Solar energy harvested by photosynthesizing organisms drives almost every food web on land and in the ocean. The marine environment is divided on the basis of the availability of light into two main zones: the photic zone (the sunlit part of the marine environment which receives enough light for photosynthesis to occur) and the aphotic zone (the inky dark part of the marine environment where photosynthesis does not take place).

The part of the solar spectrum used for photosynthesis is called Photosynthetically Active Radiation (PAR). It equates more or less to the visible spectrum and includes wavelengths from 400 nm (blue light) to 700 nm (red light). Only visible light (and part of the ultraviolet) is transmitted through seawater to any significant depth.

In marine science light intensity is commonly measured as irradiance. This is the amount of energy falling on a unit surface area in unit time. Units vary, but for biological purposes it's commonly measured as photons per square metre per second.

Photons that make up the PAR pass through the atmosphere relatively unimpeded. But when they reach the surface of the sea they may be reflected before they even penetrate into the sea. Those that do manage to penetrate are affected by the water molecules and anything else they

Figure 1.9 Light penetration in the euphotic zone.

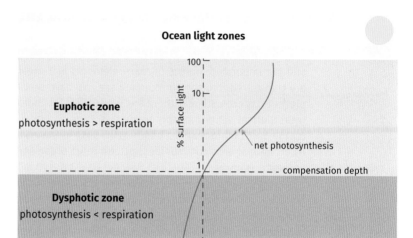

hit. Typically, PAR decreases exponentially with depth. Even in the clearest ocean waters, sunlight that can be used by photosynthesizing organisms rarely penetrates more than 200 metres.

Functionally, the photic zone is divided into the euphotic zone or 'sunlit zone' and the dysphotic zone or 'twilight zone'. The euphotic zone is commonly defined as the part of the marine environment in which there is enough light for photosynthesis to exceed respiration and for growth and reproduction to occur. The dysphotic zone is that part of the marine environment where there is enough light for photosynthesis to take place, but at a slower rate than respiration (see Figure 1.9).

Defined in this way, the depth of the euphotic zone equals the depth at which irradiance is at the compensation point (the point at which photosynthesis and respiration are in balance). However, compensation points are species specific, and can vary with time and location even within species. For practical purposes, therefore, the depth of the euphotic zone is usually taken as the point at which only 1 per cent of the surface PAR remains. Measured in this way, the depth of the euphotic zone is an indicator of water clarity. Although this is an important property for primary production and heat transfer in the upper water column, it is not a measure of photosynthetic activity.

The depth of the euphotic zone varies from only a few metres in plankton-rich coastal waters to around 200 metres in the clear open ocean.

In the dysphotic zone, there's enough light for visual predators to see, and even for some photosynthesis to take place. However, the rate of respiration is greater than the rate of photosynthesis. In clear water, the dysphotic zone may extend as deep as 1000 m. Beneath the dysphotic zone lies the aphotic zone.

Hardly a photon shines on permanent residents in the aphotic zone. However, these organisms are not in perpetual darkness. Incredible underwater

Figure 1.10 Bioluminescence in a deep-sea octopus lights up its whole body. Light-production is biochemical and 'cold' (it does not warm up the environment).

Photo: Digitalbalance/Shutterstock

film in documentaries such as *The Blue Planet* have shown that deep-sea animals produce their own light and live in a wonderful world of bioluminescence (see Figure 1.10).

Bioluminescence refers to the ability of a living organism to create and emit light. It is a phenomenon that occurs in many marine organisms that live at all depths of the sea, but it is seen to greatest effect in deep-sea organisms. Marine scientists estimate that 90 per cent of animals living in waters below 500 metres are bioluminescent.

Animals, such as deep-sea octopus, which can move freely up and down the water column, can feed on the rich pickings of fish and other prey in the photic layers of the ocean. Those living permanently in the aphotic zone usually have to eat each other or catch edible material raining down from above as marine snow.

Occasionally, deep-sea animals are treated to a feast of food when the body of a dead whale sinks and makes its way down to the sea floor. A decomposing carcass of a whale on the ocean floor is known as whale fall.

A closer look at whale fall

Although dead whales are regularly washed up on beaches, an estimated 99 per cent die at sea and sink to provide an underwater feast in the deep ocean, a marine environment that tends to be food-limited. Because of the enormous body size of an adult whale and its high lipid content in both blubber and bone, a whale fall is usually a site of intense and persistent feeding activity. Dr Nicholas Higgs, a whale fall researcher, estimates that one carcass can provide the equivalent of 200 years' worth of food in the deep sea.

A whale carcass typically attracts a diverse array of creatures and becomes a significant, energy-rich ecosystem with its own particular

Figure 1.11 Zombie worms (*Osedax roseus*) living in the bones of a dead whale that has fallen to the seafloor. These worms have no mouth, gut, or hard parts. They use fleshy root-like tissues to bore their way into the bones and, with the aid of symbiotic bacteria, obtain nutrition from lipids in the bone marrow.

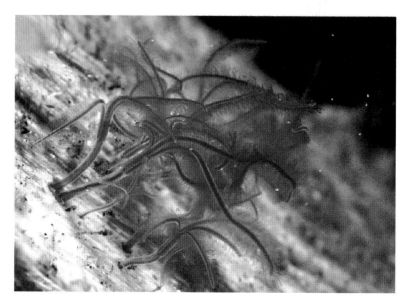

Photo: Yoshihiro Fujiwara/JAMSTEC

(and very peculiar!) community. More than 120 new species have been discovered living in whale falls, including *Osedax*, bone-eating polychaetes that have acquired the nickname 'zombie worms' (see Figure 1.11).

As a whale carcass sinks and then settles on the seafloor, it goes through a succession of ecological stages characterized by different types of organisms. A large whale carcass can act as a source of food for many years.

There are four main stages of ecological succession in a whale fall, with each stage bringing new organisms to the whale feast:

1. **The mobile-scavenger stage:** this is the first stage, when active scavengers such as sleeper sharks, hagfish, and (when the carcass is near the sea surface) seabirds consume the soft tissues (duration up to 2 years);

2. **The enrichment-opportunist stage:** this is when huge numbers of invertebrates, such as giant isopods (woodlouse-like sea creatures), feed on blubber and other organic matter remaining on the carcass (duration up to 2 years);

3. **The 'sulfophilic' (or 'sulfur-loving') stage:** in this stage, anaerobic chemosynthetic bacteria decompose bone lipids and release hydrogen sulfide that can be used by other bacteria, both free-living forms and those living in symbiosis with animals. Sulfophilic bacteria provide a source of food for a community that includes bone-eating worms (Figure 1.11), bivalve molluscs, and crustaceans (duration up to 50 years);

4. **The reef stage:** here most of the tissue has been broken down and the skeleton functions simply as a hard substrate reef for suspension feeders.

As with all ecological successions, a whale fall community changes continuously and the stages overlap. The duration and overlap of stages can vary markedly with the size of the whale carcass and seafloor environment.

Scientific approach 1.3
Uncovering the mysteries of marine snow

Marine snow includes carbon- and nitrogen-containing bits of dead animals and algae, faeces, and decaying matter that fall from the surface and middle of the water column toward the seafloor. These bits acquired their name

Figure A A robotic ocean profiler being dropped into the sea to study fragmentation of marine snow.

Photo: Giorgio Dall'Olmo and Bob Brewin

because when ocean scientists descended through the water column in a submersible, the bits made it look as if they were moving through a snowstorm.

Marine snow contains sufficient energy and nutrients to sustain large numbers of marine organisms in the aphotic zone. Unconsumed marine snow eventually sinks to the bottom where it becomes incorporated into the layer of muddy ooze that forms a thick blanket over much of the seafloor. Marine snow that sinks to the bottom of the deep ocean can store carbon there for hundreds of years, or longer.

In 2020, scientists from the National Oceanography Centre in Southampton, Plymouth Marine Laboratory, and France's National Centre for Scientific Research reported that around 70 per cent of the marine snow formed at the surface is lost before it even reaches the aphotic zone. Around half of this 70 per cent was thought to be consumed by zooplankton and microorganisms, such as bacteria, but the fate of the other half was unclear.

Marine snow is defined as consisting of bits more than 5 mm in diameter. Much of it is rich in mucus and therefore individual bits tend to grow by sticking together to form larger bits. The scientists hypothesized that a significant proportion of marine snow was fragmenting into smaller particles rather than growing into larger flakes. Fragmentation could prevent the carbon-containing matter in marine snow from reaching the sea floor of the deep ocean. However, the scientists could not test their hypothesis until they had a means of measuring the fragmentation. They did this by adapting battery-operated autonomous devices, called robotic ocean profilers (Figure A). These profilers were designed as key components of the Biogeochemical-Argo project to collect high quality temperature and salinity data from the top 2000 metres of the ocean.

Twenty-four profilers were equipped with optical sensors to detect both marine snow and smaller, non-sinking particles. Moving up and down in the ocean, the profilers recorded thousands of particle measurements that enabled the scientists to compare the loss of marine snow particles with the appearance of smaller fragments at different depths. The measurements supported the hypothesis that particles were indeed fragmenting at a high rate, roughly equal to the previously measured particle consumption processes.

These observations indicated that fragmentation may be the most important process stopping **carbon sequestration** by marine snow.

❓ Pause for thought

Suggest possible reasons for the fragmentation of marine snow.

Why is the sequestration of carbon in the sea floor of the deep ocean so important?

Chapter summary

- The five major oceans on Earth interconnect to form one huge body of seawater called the World Ocean.
- Globally, the marine environment refers to all the abiotic and biotic conditions that affect the World Ocean.
- The marine environment can be divided into polar, temperate, and tropical zones.
- The polar zones occur in the Arctic and Antarctic; ice algae form the base of polar food webs, with krill playing a central role in the Antarctic.
- The temperate zones are characterized by seasonal fluctuations in weather and primary production.
- The warm sunlit waters of the tropical zones have a very low biodiversity in the open ocean and a very rich biodiversity in coral reefs.
- Spatially, the marine environment can be divided into two major zones, the pelagic and benthic zones.
- The pelagic zone refers to the open ocean and has two main groups of organisms: swimmers (nekton) such as dolphins, and drifters (plankton) which range in size from viruses to large jellyfish.
- The benthos refers to the sea floor which contains many ecosystems including hydrothermal vents and seamounts. It ranges in depth from the intertidal zone to the hadal zone.
- Food availability in the deep sea is usually poor but is enriched by marine snow and whale fall.
- The marine environment can also be divided into a sunlit photic zone and a dark aphotic zone.
- The photic zone is divided into the euphotic and dysphotic zones, based on changes of light intensity with depth and the compensation depth.
- The aphotic zone is not completely dark due to the bioluminescence of many of its inhabitants.

Further reading

Danarvo, R., Snelgrove, P.V.R., and Tyler, P. (2014) Challenging the paradigms of deep-sea ecology. *Trends in Ecology & Evolution*, 29, 465–475.
Available on-line as a pdf. A fascinating review that describes new technologies and discusses the importance of symbioses in deep sea ecosystems.

Hardy, A.C. (1967) *Great Waters*. Collins.
Inspirational account of life as a marine zoologist on a research vessel in the Antarctic.

Mladenov, P.V. (2013) *Marine Biology: a very short introduction*. OUP.

An expert and concise introduction to marine ecosystems from shores to the deep sea.

Stow, D. (2017) *Oceans: A very short introduction*. OUP.

An accessible, interesting, and inexpensive introduction to the science of the sea, elegantly written by a Professor of Oceanography.

Thomas, D.N, and Bowers, D.G. (2012) *Introducing Oceanography*. Dunedin Academic Press.

A readable, well-illustrated, and succinct overview of oceanography aimed at students and non-specialists.

 Discussion questions

1.1   News reports and social media often refer to the threats of pollution and global climate change to the marine environment. Discuss the different understandings that can arise when people use a seemingly simple term such as 'the marine environment'.

1.2   Suggest reasons why scientists have devised a generally accepted classification of the terrestrial environment into biomes, but this has not been possible for the marine environment.

1.3   In *Challenging the paradigms of deep-sea ecology* (see Further reading), Danarvo and his co-authors refer to '... new ecological interactions and the importance of "dark energy" ... in fuelling biodiversity' in the deep sea. What do you understand by this 'dark energy'?

2 THE OCEAN IN MOTION

Inspired by his famous relative Sir Ernest Shackleton, Keith Shackleton became renowned for depicting the unforgiving icy seas around the Arctic and Antarctic. His painting *Rough seas with albatross in flight* (Figure 2.1) gives us a glimpse of what it can be like to be above an ocean in motion.

Figure 2.1 *Rough seas with albatross in flight.* A copy of an oil painting by Keith Shackleton.

© Rountree Tryon Gallery

The Southern Ocean is the most treacherous of all the seas. With no land to stop them, its winds whip up waves which are tens of metres high. It might seem impossible that any creatures could survive such conditions, but they do. And probably the greatest of them all is the Wandering Albatross, *Diomedea exulans*.

Seen on land, this bird does not appear to be built for flight. With its dumpy-looking body and large webbed feet, it takes off with about as much grace as a swan in wellington boots. But once in the air, its legs safely tucked out of the way and its wings extended and locked in place, it becomes a master of the air. It does not fight the elements. It uses them.

By attaching GPS recorders to albatrosses, scientists found that they use a technique called **dynamic soaring** to take advantage of winds forced upwards by giant waves (Figure 2.2).

Dynamic soaring enables the Wandering Albatross to fly at speeds up to 20 metres per second without flapping its wings. By using this method repeatedly, it can glide for hours and travel thousands of miles without rest above the harsh marine environment that is its home.

Figure 2.2 Schematic diagram showing an albatross flying in an across-wind direction using an S-shaped dynamic soaring manoeuvre. The bird extracts energy from the wind by heading upwind and climbing, and then turning to head downwind and descending.

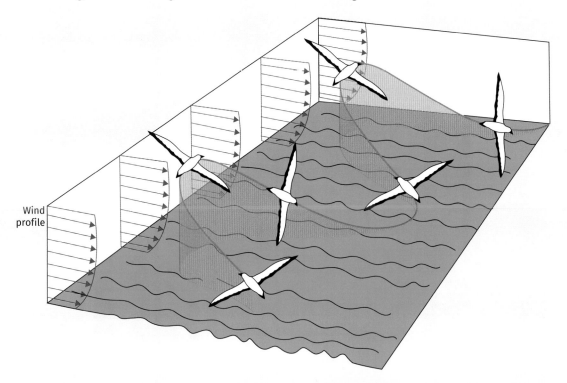

Wind profile

Source: Adapted from Figure 1 in Richardson Philip L. 2019 Leonardo da Vinci's discovery of the dynamic soaring by birds in wind shear. *Notes Rec.*73285–301 http://doi.org/10.1098/rsnr.2018.0024

The water cycle

Seawater is constantly on the move; our Ocean is a restless one. The most obvious signs of water movement are waves, the ebb and flow of tides, and surface currents. But even on a calm day with not a ripple in sight, there is movement above and below the sea surface.

Above the surface, vaporized water molecules escape into the atmosphere to play their part in the water cycle and eventually return to the ocean (Figure 2.3).

The Ocean holds 97 per cent of the total water on Earth. More than 75 per cent of global precipitation occurs over the Ocean, and it provides more than 85 per cent of the evaporated water that goes into the water cycle. Everyone, even someone living in Dzurgistan Bay in northwest China, which is the piece of land furthest (2648 kilometres) from the sea in the world, depends on the Ocean as their ultimate source of water.

Figure 2.3 The water cycle.

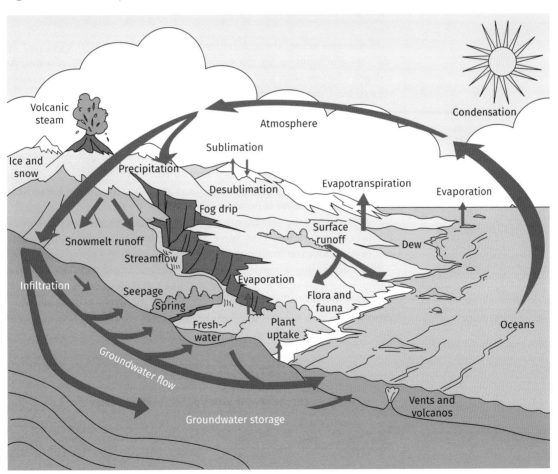

Image: John Evans and Howard Periman, USGS

Waves

Almost all waves on the surface of the sea are created by the action of wind on water. A wave might appear to be seawater travelling from one place to another, but this is not what happens. When wind disturbs water to generate a wave, individual water molecules move in an approximate circular motion with a cog-like interplay between water molecules at different depths (Figure 2.4). This explains why a seagull bobs up and down on the sea and moves only slightly forwards as a wave passes.

Figure 2.4 The action of waves in the open ocean.

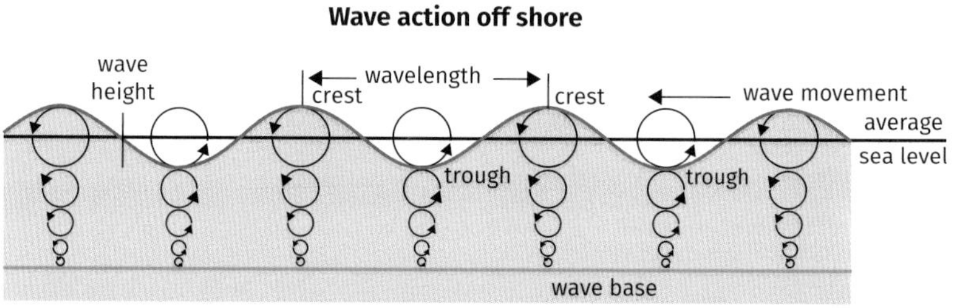

Wave action off shore

In the open ocean, the circular movements of water molecules caused by the action of wind on the sea surface decrease with the depth of water. At a depth of more than half a wavelength (the distance between successive crests), wind-generated water movement is negligible.

Major storms can produce waves powerful and durable enough to cross entire oceans. Storm waves originating in the Antarctic Ocean, and avoiding any intervening islands, are known to travel many thousands of miles and be felt on the shores of Alaska. The size and energy content of the waves are diminished as they expend energy through movement, but the waves retain the same pattern of movement they had in the Antarctic.

The speed of a wave reflects the speed of energy transfer. The bigger the wave, the greater the amount of energy transfer, and the greater the amount of energy released when the wave breaks. Individual storms can generate waves that can exert pressures of up to 30 000 kilograms per square metre.

It is usually only when sea waves reach shallow water that their potential energy is fully released. At a depth of less than half their wavelength, frictional resistance with the substrate causes the length of the waves to decrease. Wave crests come closer together and their height increases until they become so tall and unstable that they break and plunge forward as surf (Figure 2.5).

Figure 2.5 Change in the action of waves near a beach.

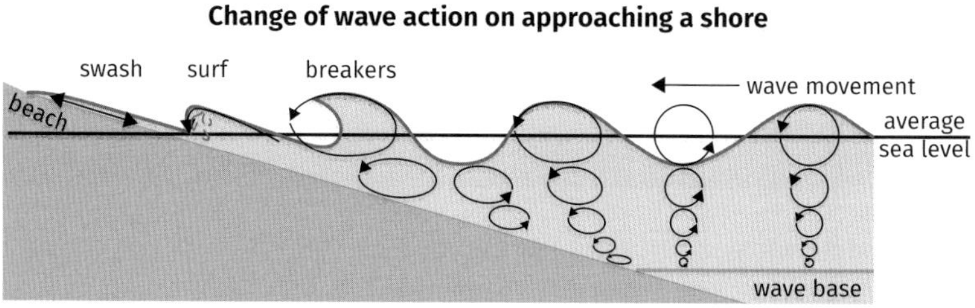

Change of wave action on approaching a shore

Wave size depends mainly on wind strength, duration, and fetch (the uninterrupted distance over which the wind is blowing). A storm blowing across the Atlantic Ocean will generate much larger waves than a storm of similar strength and duration blowing in the English Channel.

Waves are one of the most important factors determining the type and diversity of organisms on a shore. A shore on a headland jutting out into the Atlantic is potentially much more exposed to wave action than a shore on the landward side of an island. But the amount of wave action experienced on a particular shore depends not only on its fetch, but also on its topography. With increasing steepness, water movement generally tends to become more violent. The wave action experienced by an organism on a rocky shore consisting of irregular slopes, platforms, and ledges, will depend on the organism's precise location.

Case study 2.1
Adaptations to a life on wave-swept rocks

Limpets are among the best known of all the animals living on seashores. The Common Limpet (*Patella vulgata*, see Figure B) is found from Norway to the Mediterranean on nearly all rocky shores, but it is most abundant on wave-swept rocks (e.g., Figure A). The high velocity of water over these rocks exerts huge hydrodynamic forces that would dislodge the limpets if they were not well adapted to their environment.

Figure A Wave-swept rocks on the north coast of Cornwall.

Photo: © Michael Kent

Figure B Common limpets, *Patella vulgata*.

Photo: © Michael Kent

The adaptations of the limpet include its ability to glue itself onto rocks and the resistance of its shell to the hydrodynamic forces acting on it.

The limpet's almost proverbial powers of adhesion are commonly thought to result mainly from suction. This is reflected wonderfully in a poem by Pam Ayres which starts with the following verse:

> I am Clamp the Mighty Limpet
> I am solid, I am stuck.
> I am welded to the rockface
> With my superhuman suck.
> I live along the waterline
> And in the dreary caves.
> I am Clamp the Mighty Limpet!
> I am Ruler of the Waves.

© Pam Ayres—printed with permission, from Pam Ayres' book, THE WORKS (Ebury/BBC Books)

However, a limpet's adhesive powers are far greater than could be produced by suction alone. They are due to a combination of adaptations. First and probably foremost is the ability of limpets to secrete mucus that can act as a glue, bonding the foot of the limpet to the rock. Interestingly, the physical properties of the mucus change when a limpet moves: the mucus becomes more fluid, enabling the limpet to move speedily (well, speedily for a snail!) from one place to another. In addition, the limpet has special muscles that clamp its shell tightly down onto the rock, and it is able to grow a shell that closely conforms to the contours of the rock on which it is attached. On shores with high exposures to wave action, and on rocks soft enough to allow it, a limpet typically etches a home scar, a patch in the rock that exactly fits the shape of its shell.

It is generally believed that the conical shape of a limpet shell has evolved to minimize the large hydrodynamic forces (lift and drag) that are exerted on it. Mark W. Denny, an expert in the biology and mechanics of wave-swept environments, used limpet-like models to explore how lift and drag actually vary with the shape of the shell. Expressing shell shape as the ratio of height to radius, he found that the risk of dislodgement is minimized when the ratio is 1.06 and the apex is in the centre of the shell. However, limpets in the marine environment are seldom optimally shaped and typically have a height-to-radius ratio of 0.68 with the apex to the anterior of the centre of the shell. Denny thought that the difference between the typical and optimal shapes might be due to the extremely high adhesive powers of the limpets. He suggests that most limpets adhere to the rock so tightly that they are unlikely to be dislodged regardless of the shape of their shells.

In contrast to the limpet, the Celtic sea slug (*Onchidella celtica*) has adopted a completely different strategy for coping with living in a wave-swept environment. Instead of resisting the high hydrodynamic forces exerted by powerful waves, it avoids them.

The Celtic sea slug is an air-breathing gastropod that takes refuge in rock crevices in the intertidal zone. It emerges only on an ebbing tide when the rocks are fully exposed to air. Its powers of adhesion are so weak that it can be dislodged by a strong gust of wind.

At low tide on some rocky shores, such as those on the north coast of Cornwall (Figure A), Celtic sea slugs and limpets can be seen feeding alongside each other on the open rock. They both use a **radula** (a toothed,

Figure C A Celtic sea slug (*Onchidella celtica*), feeding on sea lettuce (*Ulva lactuca*).

Photo: © Michael Kent

ribbon-like feeding structure found in most gastropods) to scrape organisms off the rock surface. And they both undertake foraging excursions from their home site to which they return when foraging is finished.

❓ Pause for thought

Denny concluded that the evolution of a 'tenacious adhesion system' in limpets '... pre-empted the selection for a hydrodynamically optimal shell, allowing the shell to respond to alternative selective factors'. To what other 'selective factors' might the shape of the shell be responding?

The ability of Celtic Sea slugs to live on wave swept shores depends on them being able to take refuge in their crevice. If they fail to reach their home site before the tide returns, they are likely to be swept into the open ocean. Homing occurs in limpets, but it is not such a life-and-death affair. Limpets that fail to reach their home site before the flood tide reaches them are unlikely to be swept away. Suggest why homing behaviour has evolved in limpets—how does homing benefit them?

Wave-causing submarine disturbances

Waves are not only generated by the wind acting on the sea surface, they can also result from submarine disturbances. These disturbances include explosions caused by volcanic activity and earthquakes, and sediments sliding and slipping in the sea. Small disturbances are common and usually have only local significance. But large disturbances can lead to the generation of giant waves called tsunamis that devastate basin-wide areas.

Tsunamis destroy ecosystems, especially those on steep slopes in shallow seas and in the intertidal zone where the forces of a tsunami become more concentrated. Coral reef, mangrove swamp, and sea grass ecosystems are particularly vulnerable to tsunamis. Tsunamis can also have catastrophic effects on humans. For example, an exceedingly powerful tsunami, generated by a combination of explosive volcanic activity and sediment slides, is believed to have destroyed the Minoan civilization on Crete some 3000 years ago.

Tides

Tides, the rhythmical rise and fall of sea level, cause the movement of enormous masses of seawater in the intertidal zone. In marine biology, the flow of water towards the land is called the flood tide, and the flow of water away from the land is called the ebb tide.

Tides cause an intertidal zone to be completely immersed (submerged by seawater) at the highest high tides and completely emersed (exposed to the air) at the lowest low tides (Figure 2.6). Organisms at the highest level live in an incredibly variable environment. They are exposed to air for prolonged periods and experience extremes of temperature and salinity.

Figure 2.6 The same beach looks very different at high and low tides.

(a)

(b)

Photos: © Anthony Short

Case study 2.2
A closer look at life between the tides

Rachel Carson, in her beautifully written book *The Restless Sea*, sums up brilliantly the advantages and disadvantages of living between the tides. She writes:

> The billions upon billions of sessile animals, like oysters, mussels, and barnacles, owe their very existence to the sweep of the tides, which brings them the food which they are unable to go in search of. By marvellous adaptations of form and structure, the inhabitants of the world between the tide lines are enabled to live in a zone where the danger of being dried up is matched by the danger of being washed away, where for every enemy that comes by sea there is another that comes by land, and where the most delicate of living tissues must somehow withstand the assault of storm waves that have the power to shift tons of rock or crack the hardest granite.

Not everyone can venture into the deeps and experience for themselves the amazing and peculiar organisms that inhabit hydrothermal vents, nor swim with manta rays along the Great Barrier Reef. But many of us at some time can go down to a seashore and encounter some of the marine organisms that Rachel Carson writes about (Figure A).

Any budding marine zoologist might also be encouraged to visit a shore by the words of C.M. Yonge. In his classic New Naturalist book, *The Seashore*, he writes:

> The seashore is the meeting place of sea and land. It is for that reason the most fascinating and most complex of all the environments of life. It has long been recognised as the best training ground for a zoologist because upon it are to be found members of almost all of the invertebrate groups of animals and with them shore fishes and even, on occasion, marine mammals, such as stranded whales or breeding seals.

Figure A A tidal pool, just one of many marine habitats between the tides.

Photo: © Michael Kent

Figure B A schematic of a rocky shore in the British Isles showing the basic 'zones' and how some environmental conditions change with height up the shore.

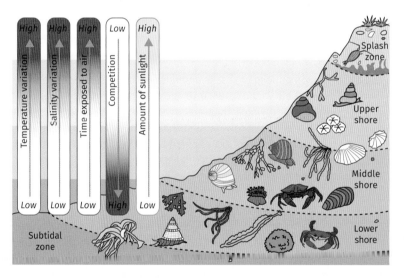

Adapted from a drawing by Jack Sewell for the Marine Biological Association, UK.

❓ Pause for thought

Figure B shows some of the environmental conditions that change with height up a shore. What other conditions are likely to vary progressively up the shore? Where do you think there will be the greatest biodiversity?

What causes tides?

Observing tides is easy. Just go down to a shore, set up camp to spend a day looking at the water's edge and you will see the tide move up and down the shore. You could even use an automated device to measure and record the relative height of the tide. On most shores in the world, if you did this for every hour of every day over a whole year, you couldn't help but notice the following:

- there are two high tides and two low tides, on average every 24 hours 50 minutes
- the tidal range varies with the phase of the Moon
- the highest high tide and lowest low tide occur around the spring and autumnal equinoxes.

From these observations, you might surmise that the behaviour of the Moon and Sun has some type of association with tides. And you'd be right. However, this does not explain what causes tides. It took the genius of Sir Isaac Newton to show that tides are just one of the many phenomena in the Universe that are driven by the gravitational forces of attraction that exist between any two bodies that have mass.

A detailed mathematical explanation of tides is beyond the scope of this book. But for most marine biological purposes it's sufficient to know that tides are mechanical phenomena driven mainly by gravitational forces of attraction between the Moon, the Sun, and a rotating Earth (Figure 2.7).

Figure 2.7 Cause of spring and neap tides. When the Earth, Moon, and Sun are in line (left), gravitational forces from the Sun and Moon combine to generate tides higher and lower than normal (spring tides). In contrast, when the Moon is not in line with the Sun and Earth (right), the gravitational force of the Moon acts in opposition to that of the Sun, generating moderate tides (neap tides).

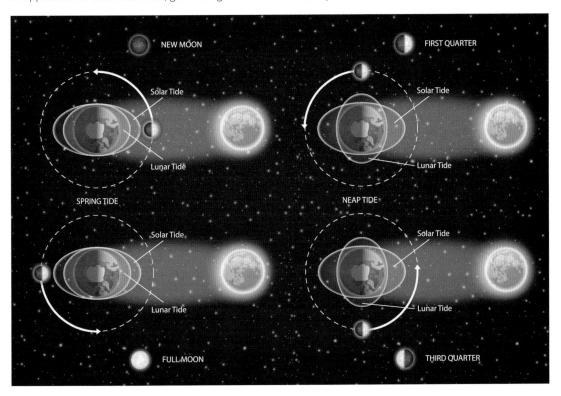

Image: Siberian Art/Shutterstock

The combined tide-generating forces of the Sun and Moon depend on their positions in relation to each other and the Earth. When the Sun, Moon, and Earth are in alignment at full moon and new moon, the forces are at their greatest. This leads to big tides called spring tides.

When the gravitational forces of the Sun and Moon are at right angles to each other (i.e., just after the first quarter and third quarter of the moon), the tidal range (the difference between the height of the sea at low tide and at high tide) will be least. These are called neap tides.

The tidal range becomes progressively greater towards a spring tide, and progressively smaller towards a neap tide.

Tidal range varies seasonally as well as daily and during a lunar cycle. This is because the Earth moves elliptically around the Sun approximately once every year. The Sun exerts its greatest tide-generating force around the spring and autumnal equinoxes when it is closest to the Earth. Therefore, the spring tides in March and September usually have the highest tidal ranges of the year. These equinoctial spring tides enable a marine biologist to explore the lowest part of the shore without getting wet.

The interplay between the gravitational forces of the Sun and Moon and a rotating Earth are complicated by the fact that seawater covers less than three-quarters of the Earth's surface and varies in depth. Land interrupts the flow of tidal water movements. The gravitational pull of the Sun and Moon sets up complicated oscillations of seawater in each ocean basin; this has been likened to water sloshing about in a bowl. Tides are modified greatly by landmasses, the shape of each basin, and the topography of particular shorelines. Consequently tidal ranges and tidal patterns vary from one location to another.

In the open ocean, the tidal range is generally no more than 1 metre. The greatest tidal range occurs in the Bay of Fundy, Newfoundland, where extreme low spring tidal heights are 16 metres below extreme high spring tidal heights. In contrast, in the semi-enclosed Mediterranean Sea the tidal range is typically less than 15 centimetres.

Although most localities have two low tides and two high tides of the same height each lunar day (called semi-diurnal tides), some places have two high and two low tides of different heights (mixed semi-diurnal tides), and a few places have only one high and one low tide each day (diurnal tides). There is also a locality in the Ocean, called the amphidromic point, where there is no tide.

Tidal predictions

Today, computer models incorporating the most up-to-date mathematical formulations of tides and the effects of local factors are used to predict tides at a large number of ports throughout the world. These predictions are available in various forms of 'tide tables'.

Anyone planning to carry out an intertidal investigation should consult the most appropriate tide table. Knowledge of the predicted tides is essential for working safely and effectively on the shore. However, it is important to be aware that tide tables contain predictions only for standard meteorological conditions. Changes in weather can affect these predictions. Variations in atmospheric pressure can cause tides to be as much as 0.3

metres different from those predicted. A low atmospheric pressure acting on the sea surface will tend to allow sea levels to rise more than normal; a high atmospheric pressure acting on the sea surface will to tend to cause tides to be lower than predicted. Onshore winds can cause water to pile up, resulting in tides being higher than predicted. Storm surges, caused by a combination of high onshore winds and low atmospheric pressure, can raise sea levels by one metre or even more, making the shore much more dangerous to work on during a flood tide. Therefore, before working on the shore it's prudent to consult the weather forecast as well as tide tables.

Tidal currents

Horizontal movements of seawater associated with tides are called tidal currents or tidal streams. Tidal currents are usually weak in the open ocean, but they can be extremely strong where water is funnelled through narrow passages. They occur in the sublittoral zone as well as in the intertidal zone.

Strong sublittoral tidal currents can be rich in nutrients and plankton which support high levels of biodiversity. For example, very strong tidal currents in the Menai Straits (north Wales) support a very diverse community of sessile (fixed) organisms, including many species of filter feeding hydroids and sea anemones.

Case study 2.3
Grunion: A fish out of water

The California Grunion (Figure A) is a very peculiar fish indeed: it mates completely out of water and its offspring spend the first couple of weeks of life buried in sand.

The breeding cycle of the California Grunion is exquisitely timed to take full advantage of the tides and the particular environmental conditions on the surf-swept shores of south California. At night during March to August around a full or new moon and just after high tide, grunion can be seen in their thousands wriggling and writhing on the wet sand above high tide. They have got there by riding on the crests of large waves and flinging themselves onto the wet sand above the reach of the high tide. Males and females intertwine, eggs (about 3000 from each female) are fertilized and laid in a shallow depression in the sand (excavated by the female using her tail), and then the parents head down the beach and hitch a ride back to the sea on an ebb tide. Abandoned by their parents, the embryos are left to develop buried in sand above the reach of tides for the whole of their incubation period. The eggs remain in the sand until the next batch of spring tides brings surf high enough to reach them and trigger hatching. This is usually about 11 days after the eggs are laid. Released from the confines of the egg case, young grunion are washed out to sea where they lead a typical free-swimming life as part of the nekton.

Figure A Egg laying and hatching in the California Grunion (*Leuresthes tenuis*).

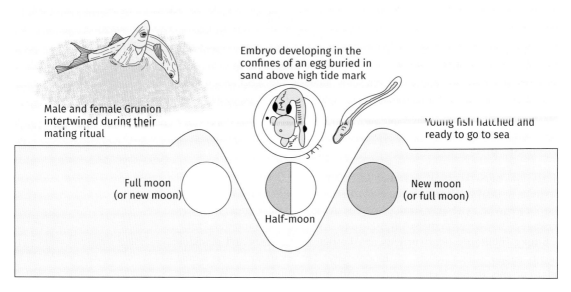

Embryo developing in the confines of an egg buried in sand above high tide mark

Male and female Grunion intertwined during their mating ritual

Young fish hatched and ready to go to sea

Full moon (or new moon)

Half-moon

New moon (or full moon)

Source: Introduction to grunion biology/Karen Martin, 2006. http://grunion.pepperdine.edu/ IntroductionToGrunionBiology.pdf Reproduced with permission from artist, Greg Martin.

Each stage of the grunion's breeding cycle is finely tuned to the rhythm of the tides and the phases of the Moon. But it is unclear whether the Moon acts directly on the grunion or indirectly through the tides it generates. As grunion runs occur at new as well as full moon, it cannot be the light of a waxing moon that drives grunion into a breeding frenzy. And there is no scientific evidence that grunion can detect variations in the gravitational pull of the Moon. Professor Ernest Naylor, an expert in the rhythmic behaviour of marine organisms, concluded that '... *a reasonable working hypothesis is that its behaviour is under biological-clock control and is phased by the tides rather than directly by the Moon*'.

A **biological clock** is an internal 'molecular time piece' controlling periodic processes such as reproductive behaviour. The clock controlling the grunion run would be a circa-semilunar 14.75 day clock reset to the local lunar cycle experienced by grunion before they mate on a Californian shore. Circa-semilunar rhythms and clocks are widely present in marine organisms, but researchers have identified the precise molecular basis of only a few.

The California Grunion has probably the most unusual reproductive behaviour of any fish, but we still have much to learn about how it times its mating so immaculately to the phases of the Moon.

❓ Pause for thought

Most biological adaptations are trade-offs between competing effects. For the California Grunion, the benefits of laying eggs in the sand above the normal high tide mark must outweigh the costs. What are the advantages and disadvantages of a marine fish adopting this breeding strategy?

The ocean circulation system

Surface currents

Ocean currents are directional movements of enormous masses of seawater. Surface currents are those in the upper ocean layer (Figure 2.8). They are driven mainly by prevailing winds (the most common and persistent winds in different regions of the world). A wind moving at 50 kilometres per hour for a prolonged period can produce a current of about 2 kilometres per hour.

The current in the Gulf Stream is extremely fast. It averages about 4 kilometres per hour along the eastern coast of North America. After flowing off the North American coast, the warm waters of the Gulf Stream are deflected eastwards and continue as the North Atlantic Drift; this bathes the shores of south west Britain and makes its winters relatively mild.

Over periods of months to years, prevailing winds set up an interconnecting series of almost circular systems called gyres. At the centre of each ocean gyre, there is usually a large area of relatively stagnant, clear water. In the North Atlantic, within the gyre created by the Gulf Stream, North Atlantic Drift, and the Canary and North Equatorial Currents (Figure 2.9) this area is called the Sargasso Sea; it is roughly 3000 kilometres long and 1000 kilometres wide.

Figure 2.8 The average pattern of global surface currents. The strength and direction of each current varies with the seasons and from year to year.

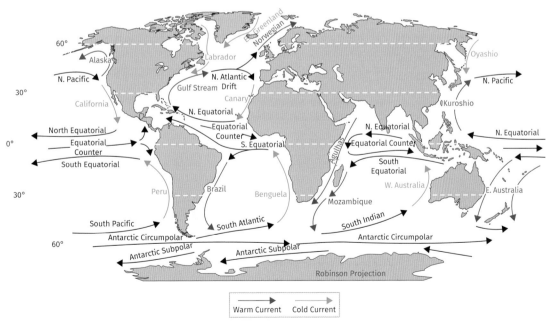

Source: Dr. Michael Pidwirny

Figure 2.9 The Sargasso Sea, located within the gyre created by the Gulf Stream, North Atlantic Current, the Canary Current, and the North Equatorial Current.

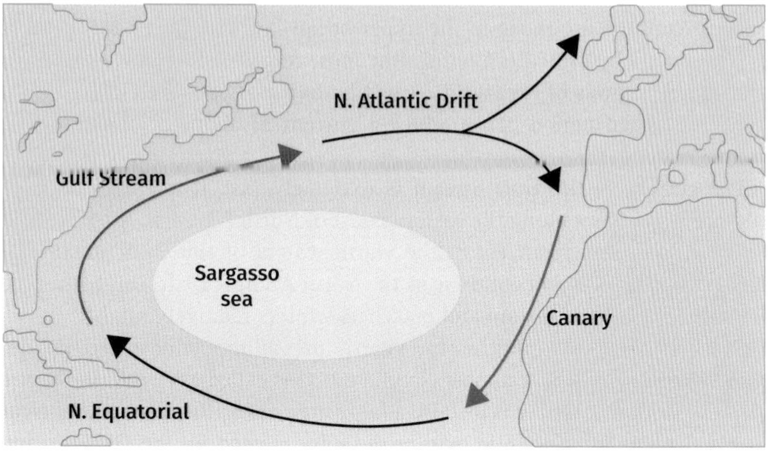

Rachel Carson called the Sargasso Sea '*a place forgotten by the winds, undisturbed by the strong flow of waters that girdle it as with a river*'. As well as being relatively calm, this tropical sea is also warm and salty. The combination of relative stillness, warmth, and saltiness has created a special marine environment with its own community of floating organisms. Chief amongst them is *Sargassum*, a seaweed with, literally, no fixed abode that is buoyed up by gas bladders and grows so profusely that it creates a floating forest.

Living among the weed are a wide variety of invertebrates along with many species of fish. Some species, such as the Sargassum Crab (*Planes minutus*) and the Sargassum Pipefish (*Syngathus pelagicus*) are endemic to the floating seaweed community. Loggerhead Turtles (*Caretta caretta*) are also reported to spend part of their first year of life under the protection of the seaweed. And seabirds such as shearwaters use large dense rafts of *Sargassum* as safe roosting sites.

While the relative stillness of the Sargasso Sea creates conditions that enable a floating forest of seaweed and its community to develop, the fast movement of the Gulf Stream in the nearby gyre enables other organisms to drift long distances and increase their feeding opportunities. One of the most prominent of these drifting organisms is the Portuguese Man-of-War, *Physalia physalis*.

A closer look at the Portuguese Man-of-War

The Portuguese Man-of-War (Figure 2.10) appears to be a single balloon-like jellyfish with a pink crest and long blue tentacles. But it is not a true jellyfish, nor, according to some zoologists, even a single individual organism.

Figure 2.10 The Portuguese Man-of-War (*Physalia physalis*).

Image courtesy of Islands in the Sea 2002, NOAA/OER. - U.S. Department of Commerce, National Oceanic and Atmospheric Administration

Jellyfish and siphonophores belong to the phylum Coelenterata (the phylum to which corals and sea anemones also belong, sometimes known as the phylum Cnidaria). Coelenterata are characterized by having two layers of cells, a single body opening which acts as a gut and anus, and specialized stinging cells called nematocysts.

Coelenterata come in two main body forms: a polyp, usually a simple sac crowned with stinging tentacles (adult sea anemones take the form of a polyp); and a medusa, that can be thought of as an upside down polyp, morphed to assume the shape of an umbrella, with the mouth and tentacles hanging downwards rather than upwards. As adults, true jellyfish are single animals which adopt the body form of a medusa. Adult siphonophores are actually colonies of polyps.

The Portuguese Man-of-War consists of medusae and three types of polyps. The medusae are asexual and are responsible for making the gas-filled bladder (called a sail or pneumatophore). They can also make the pneumatophore collapse and change shape as required. For example, the float may be deflated in stormy weather, and in strong sunshine. In these conditions, a Portuguese Man-of-War is often seen dipping the float under the water. After dipping, the float rights itself and the sail re-inflates.

One type of polyp is responsible for finding and capturing food. Its stinging cells inject venom into their prey and into any swimmer or surfer unfortunate enough to make contact. The tentacles can dangle more than 20 metres behind the bladder. The poison is usually no more painful than that of a bee sting to most humans, but it has been responsible for several mortalities. Another type of polyp is responsible for ingesting and digesting prey. And the third type of polyp is sexual and produces gametes.

Coriolis effect and Ekman spiral

The direction of a surface current roughly follows that of the prevailing wind that drives it. But the current's direction is also greatly affected by the Coriolis effect, interactions with subsurface currents, and large landmasses.

The Coriolis effect results from the Earth's rotation causing points at the equator to be spinning faster than those at the poles (for example, points on the equator are spinning at about 1600 kilometres per hour while those 160 kilometres from the poles are spinning at about 30 kilometres per hour).

The Coriolis effect causes water, wind, or any other object not attached rigidly to the Earth's surface to veer to the right (clockwise) in the northern hemisphere and to the left (counter clockwise) in the southern hemisphere. The Coriolis effect results in the Gulf Stream veering to the right in the north Atlantic, contributing to the formation of the gyre that encloses the Sargasso Sea (Figure 2.9).

Although a surface current is wind driven, the current does not move in the same direction as the wind. This is due to a phenomenon that involves a combination of drag forces and the Coriolis effect. Functionally, a surface current can be as much as 100 metres deep and consists of different layers. The uppermost layer exerts a drag force on the layer immediately below it and so on downwards until all the wind's energy is transferred to the sea. Each successive layer moves more slowly than the layer above and is deflected by the Coriolis effect. These movements create a downward spiral of seawater known as an Ekman spiral (Figure 2.11). It has been likened to a whirlpool with a very slow spin. The stronger the wind, the deeper is the spiral. The average direction of a surface current is at an angle roughly 45 degrees to the direction of the surface wind. The net transport of seawater in a direction at right angles to a prevailing wind is called Ekman transport.

In theory, water should spiral consistently anticlockwise down a plug hole in the northern hemisphere, clockwise down a plug hole in the southern hemisphere, and straight down a plug hole on the equator. Scientists have confirmed this outcome in finely controlled laboratory experiments. But the

Figure 2.11 An Ekman spiral in the northern hemisphere. Because the deeper layers of water move more slowly than the shallower layers, they tend to twist round and flow in the opposite direction to the surface current.

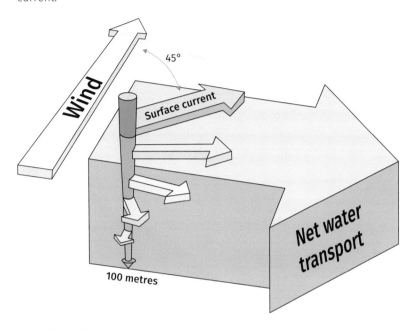

Copyright NOAA

Coriolis effect is very sensitive to other forces and water spiralling down a plughole in a typical domestic sink does not always follow the theory.

Eckman transport is responsible for upwelling, the mass movement of bottom water to the surface (Figure 2.12). It occurs when a longshore wind (a wind moving roughly parallel to the edge of a continent) drives masses of surface water offshore and is replaced by bottom water. The strength of upwelling depends on the strength of the wind and tends to be seasonal.

In seas off the British coast, the strong upwelling of cold, nutrient-rich water in spring and autumn stimulates immense algal blooms that are sources of food for many marine animals.

Subsurface currents and thermohaline circulation

Beneath surface currents, deeper water masses of the World Ocean are set in motion mainly by density changes in seawater that occur in the polar regions. This density-driven movement of the sea is called thermohaline circulation because it is due to temperature and salinity changes.

The freezing waters in parts of the Arctic off Iceland and Greenland have been described as the 'heart' of thermohaline circulation. Here warm surface water originating from the Gulf of Mexico is cooled and its salinity increased by the addition of salt resulting from the freezing of Arctic seawater (Figure 2.13). Consequently, the water becomes increasingly dense and sinks rapidly to great depths forming what is called the North Atlantic Deep Water.

Figure 2.12 Upwelling of bottom water along the west coast of a continent in the southern hemisphere bringing nutrient-rich water to the surface.

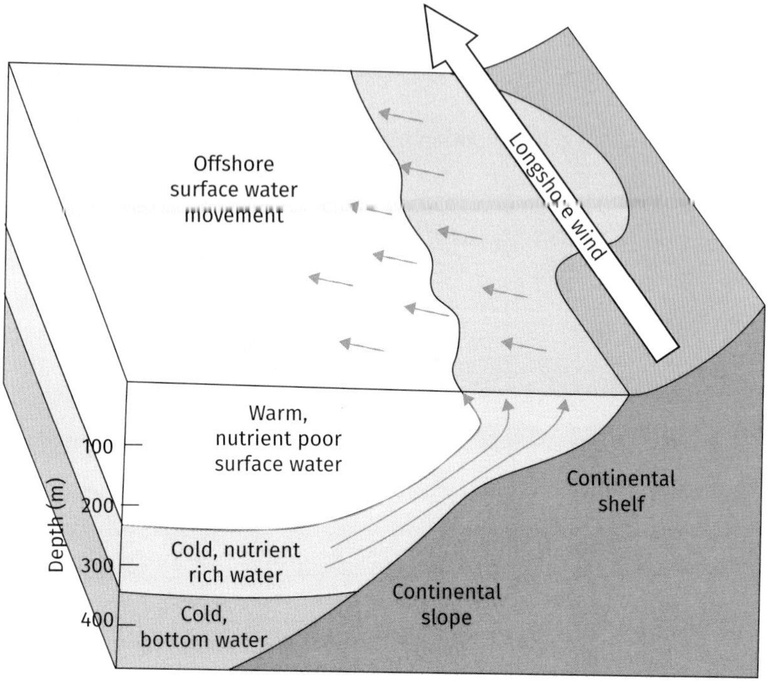

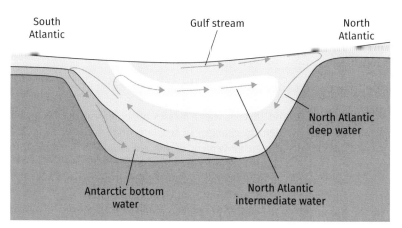

Offshore
surface water
movement

Longshore wind

Warm,
nutrient poor
surface water

Continental
shelf

Depth (m)

100
200
300
400

Cold, nutrient
rich water

Cold,
bottom water

Continental
slope

Source: Marine Biology: A very short introduction / P.V. Mladenov, 2013

Figure 2.13 Simplified schematic of the Atlantic Ocean thermohaline circulation.

South
Atlantic

Gulf stream

North
Atlantic

North Atlantic
deep water

Antarctic bottom
water

North Atlantic
intermediate water

Source: Marine Biology: A very short introduction / P.V. Mladenov, 2013

Figure 2.14 The great Ocean Conveyor Belt. Illustration depicting the circulation of the global ocean. Throughout the Atlantic Ocean, the circulation carries warm waters (red arrows) northward near the surface and cold deep waters (blue arrows) southward.

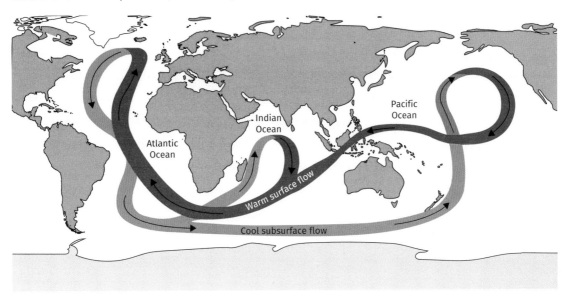

Image courtesy of NASA/JPL

This water flows near the bottom of the Atlantic and moves very slowly southward. A similar process occurs off the Antarctic. It results in a deep cold water current (called the Antarctic Bottom Water) flowing northwards. The combined movements of the warm surface currents and cold deep currents form what is known as the Atlantic Meridional Overturning Circulation. It is a key component of a remarkable natural phenomenon called the global or great Ocean Conveyor Belt that interconnects all parts of the World Ocean (Figure 2.14).

In contrast to surface currents, which can be extremely fast, deep ocean currents are very slow. They typically flow at only a few tens of metres per hour. Despite their slow speed, deep ocean currents move huge masses of water and keep the Conveyor Belt moving.

As a consequence of the deep ocean currents moving so slowly, it is estimated that it would take about 1000 years for a molecule of water to make a complete circuit of the Ocean Conveyor Belt.

The Ocean Conveyor Belt plays an enormously important role in the World Ocean and is a key component of the Earth's climate system. It transports thermal energy, living organisms, and other matter all around the world. Any change in the characteristics of the Ocean Conveyor Belt would have a global impact on marine biodiversity, and have profound consequences for human society. We will be considering these impacts in Chapter 7.

Chapter summary

- Seawater is constantly on the move, escaping into the atmosphere to play an essential role in the water cycle.
- The most obvious signs of water movement in the sea are surface waves driven mainly by the wind.
- Tides, the rhythmical rise and fall of sea level, are responsible for dramatic variations in conditions to which intertidal organisms have become adapted.
- The ocean circulation system consists of surface currents and subsurface currents. Surface currents form an interconnecting series of circulating systems called gyres.
- Subsurface currents are driven mainly by density changes in the seawater to create a thermohaline circulation.
- The combined movements of the cool subsurface and warm surface currents form the Ocean Conveyor Belt that interconnects all parts of the global ocean.

Further reading

Carson, R. (2018) *The Sea Around Us*. 3rd edition. Oxford University Press.
A new edition of a wonderfully written book. Originally published in 1951, it became one of the most influential books ever written about the marine environment.

Dipper, F. (2016) *The Marine World*. Wild Nature Press.
An encyclopaedic account of life in the marine environment.

Naylor, E. (2010) *Chronobiology of marine organisms*. Cambridge University Press.
An introduction to biological clocks that control the behaviour and physiological processes of marine organisms.

Yonge, C.M. (1950). *The Seashore*. New Naturalist, Collins.
Dated, but written by one of the greats of marine biology, still a good read.

Discussion questions

2.1 Exposure to wave action has seldom been quantified. Suggest why.

2.2 Many intertidal organisms become active at only certain states of the tide. How might they detect the state of the tide?

2.3 In January 1992 a shipping container carrying 28 800 plastic bath toys was lost overboard from a cargo vessel in the mid-Pacific Ocean. The toys were packed in sets of four, each containing a beaver, frog, turtle, and duck. The toys were carried by the Ocean Conveyor belt all over the world, some reaching British shores in 2003. Suggest why the toys did not all follow the same path and ended up at different places.

3 SEAWATER MATTERS

Our planet is a watery world with well over a billion cubic kilometres of seawater in the global ocean. One of the big questions in marine science is where all this water comes from. Geologists have strong evidence from ancient rocks that liquid water has existed at the Earth's surface for 3.8 billion years. But how the water got there is still the subject of much debate. There are two main theories. The first suggests that the Earth might have captured water from asteroids and comets which collided with it. The second suggests that water comes from the bowels of the Earth itself, and that it was released explosively in huge quantities during a period of intense volcanic activity in the Earth's history. Volcanoes continue to add water vapour as well as other gases to our atmosphere (Figure 3.1).

Volcanic water originates from deep within the Earth's mantle where it is trapped inside minerals such as ringwoodite. Geologists speculate that if all of this trapped water were released there is enough to fill the global ocean three times.

Ringwoodite is a bright blue mineral containing magnesium silicate to which water is tightly attached by hydrogen bond symmetrization, a special type of **hydrogen bonding**. In its hydrous form, ringwoodite is 1 per cent water by weight.

Ringwoodite has been found at depths of 650–700 kilometres in the layer of hot rock between the Earth's surface and its core. When mantle rocks such

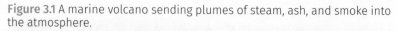

Figure 3.1 A marine volcano sending plumes of steam, ash, and smoke into the atmosphere.

Image by Julius Silver from Pixabay

as ringwoodite melt, their water liquefies into the magma. As the magma rises towards the surface and cools, pressure is reduced and the water is released and ejected as vapour through volcanoes. This process is called **degassing**.

According to the degassing theory, water vapour accumulated in the Earth's atmosphere until the planet had cooled to a temperature low enough for rain to fall. For many centuries rain fell in great torrents onto the bare rocks of the Earth's crust and drained into huge hollows to form the primeval ocean.

What makes seawater special?

Water, essential to all life as we know it, makes up the bulk of seawater with which it shares many extraordinary properties. Some of these properties result from the dipolarity of water molecules and the ability of the molecules to link together by hydrogen bonding (Figure 3.2).

Surface tension

The tendency of water molecules to bond with each other is known as cohesion. At the air–water interface on the surface of the sea, hydrogen bonds connect each individual water molecule with water molecules above and below it, but not with the air molecules in the atmosphere. The unequal distribution of hydrogen bonds produces a force called surface tension which causes seawater on the surface to contract and form a surprisingly

Figure 3.2 Dipolarity and hydrogen bonding in water molecules.
An oxygen atom has a tendency to draw electrons closer to it. This gives
the oxygen (O) end of a water molecule a slightly negative charge (δ–)
and the hydrogen (H) ends a slightly positive charge (δ+), resulting in the
tendency of water molecules to be pulled together by hydrogen bonds (1).

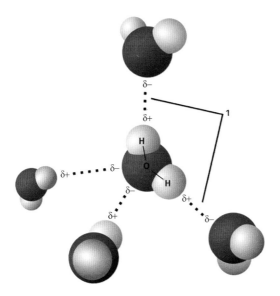

tough film or 'skin'. Surface tension is a measure of how difficult it is to
stretch or break a liquid surface. Water has a greater surface tension than
most other liquids.

Case study 3.1
Ocean-skaters (*Halobates* spp.)

The 'skin' on the surface of the sea is a habitat for many species of organism.
These include bacteria and microorganisms that live in nutrient-rich sea
foam. However, probably the most intriguing group of organisms occupying
this distinctive habitat is the insect genus *Halobates*—commonly called
Ocean-skaters (Figure A).

Five species of *Halobates* are the only insects known to live in the open
ocean. They are all relatively small—the largest adults measure only about 6mm
in length. *Halobates* species are widely distributed in the three major warm
water oceans between latitudes 40 °N and 40 °S. All five species are found in
the Pacific Ocean, but only one (*H. micans*) in the Atlantic. They generally occur
only on the surface of waters, where even winter temperatures exceed 20 °C.

Ocean-skaters are collected using specially designed tow nets that skim the
sea surface. But these insects are not easy to catch. Not only can they move

Figure A An ocean skater (*Halobates spp*)

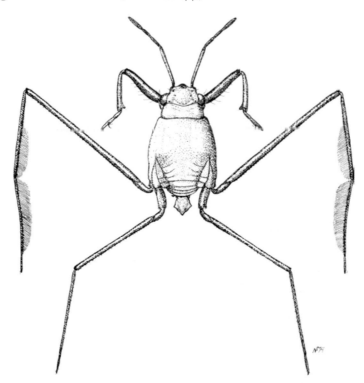

Source: Adapted from a figure in Andersen, N.M. & Polhemus, J.T. 1976. Water-striders (Hemiptera: Gerridae, Veliidae, etc.). In Marine Insects,L. Cheng (ed.). Amsterdam: North-Holland Publishing Company, pp. 187–224.

relatively quickly (up to 60 metres per minute), they are difficult to locate. Large aggregations of thousands of individuals occur in some areas, but these aggregations are sparsely distributed.

Ocean-skaters are predators that feed by sucking body fluids out of their prey. They are opportunistic feeders, catching zooplankton trapped at the sea–air interface. They have well-developed eyes which help them to find prey, locate mates, and avoid being captured.

Ocean-skaters lay their eggs on any floating object, including pieces of plastic. During one ocean expedition, 70 000 Ocean-skater eggs were discovered in an empty plastic milk jug.

Newly hatched nymphs go through five moults before reaching the adult stage, about 12 weeks after hatching, but how long Ocean-skaters live is a mystery. There remains a lot to discover about these intriguing insects of the sea.

❓ Pause for thought

Ocean-skaters are air-breathing insects that are unable to survive being under water for long periods of time. Suggest how a thin layer of tiny velvety hairs covering the body minimizes the risk to Ocean-skaters of drowning.

Thermal properties

One consequence of dipolarity and hydrogen bonding is that both seawater and pure water have a high specific heat capacity. Pure water has one of the highest heat capacities of all known substances; that of seawater is just a little less because of its salt content. Its high heat capacity means seawater has to gain a lot of energy before its temperature rises. Conversely, it has to lose a lot of energy for its temperature to fall. As a result, the temperature of seawater changes only slowly and has much lower extremes than air temperature. Consequently, most marine environments have relatively stable temperatures. Typically, ocean sea surface temperatures fluctuate less than 2 °C in any 24-hour period. But temperatures in the intertidal zone can change considerably with the ebb and flood flow of tides. Whereas most marine organisms are adapted to live within a relatively narrow temperature range, intertidal organisms have to be able to cope with diurnal and tidal changes in temperature.

Unlike many variables in the marine environment, temperature is not just a property of seawater, it is also a property of all matter, living and non-living. The internal temperature of an organism affects its metabolism and everything it does.

Seawater, like pure water, has a high latent heat of fusion as well as a high latent heat of vaporization (see Case study 3.2). This means that at its freezing point it must lose a lot of energy before it forms ice crystals. Pure water freezes at 0 °C under normal pressures, but adding salt lowers the freezing point; the more salt dissolved in the water, the lower is the freezing point. Ocean water with a salinity of about 35 grams of salt per litre only begins to freeze at about −1.9 °C.

When seawater freezes, only the water forms ice (Figure 3.3)—in fact, sea ice can be melted and used to provide drinking water. The salts are left

Figure 3.3 Icebergs are not salty—they are pure frozen water.

Photo by Jean-Christophe André from Pexels

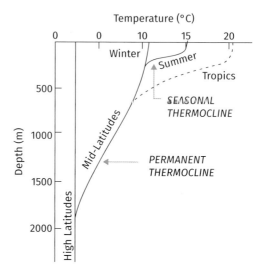

Figure 3.4 Typical seasonal temperature profile for different regions.

behind, increasing the salinity of the surrounding sea and lowering the freezing point even further.

The surface of the sea is warmed by radiant energy from the Sun. When sea temperature is higher than air temperature, thermal energy is transferred from the sea surface to the cooler air above it. Consequently, temperature fluctuations are greatest at the surface of the sea.

Surface temperatures vary with season and latitude, but fluctuations are far less than those on land. In temperate (mid-latitude) areas, sea temperatures are typically about 15 °C in the summer and about 5 °C in the winter (see Figure 3.4).

Sea temperature also declines with depth of water until it reaches a remarkably constant 4 °C in most places in the deep sea. But the decline is not usually uniform. In warm temperate and tropical seas, there is often a distinct zone of rapid temperature change, called a thermocline, between warm surface water and denser, deeper water. In tropical areas, sea temperature can exceed 25 °C immediately above the thermocline and only 6 °C immediately below it. Thermoclines are usually absent in polar regions.

Sea temperature is probably the single most important factor determining the geographical distribution of marine organisms. It has far-reaching effects on the life of every marine organism. Each marine environment has its own temperature range.

Case study 3.2
Adaptations of limpets to thermal environments

The Common limpet (*Patella vulgata*) scrapes a living on rocky shores from Norway to the Mediterranean. The thermal environment of a population depends on in which part of this geographical range it occurs. But even within each location individuals may be subjected to high levels of thermal stress. For example at Polzeath on the Cornish coast, limpets exposed to the air at low tide have to cope with temperatures that go below 0°C in winter and reach above 35°C in the summer. And on a single cloudless day in May, air temperatures ranging from 4°C to 32°C have been recorded during two consecutive low spring tides.

Common limpets generally stop feeding when the temperature falls below 10°C; they reduce their metabolism and go into a quiescent state that enables them to overwinter. The Antarctic limpet (*Patinigera polaris*) envelops its body with mucus with anti-freeze properties that enable it to survive at temperatures as low as –10°C. Although the Common limpet has no such envelope, the mucus that it uses to stick itself onto rocks might offer some protection from heat loss through conduction.

The thermal stresses to which juveniles and adults are exposed differ significantly. Juveniles tend to live low down on the shore covered in seaweed. The seaweeds not only protect limpets from solar radiation, but they also keep temperatures under the seaweeds low due to water evaporating from their damp surfaces. Adult limpets tend to live at higher tidal levels, often completely exposed to the elements.

Being **ectotherms**, limpets cannot control their internal body temperature independent of the external environment using internal, physiological control mechanisms. Their adaptations are mainly structural.

Heat transfer between the limpet and its environment takes place through various routes (Figure A). Solar radiation is the most important source of thermal energy, but heat is also gained by convection from the air and by conduction from rocks. If a limpet becomes hotter than its surroundings, it can lose heat to the air or rock. Solar radiation can be reflected and the limpet's body can be cooled by convection, and by evaporative water losses.

The colour, shape, and texture of limpet shells vary as an adaptation to heat transfer. For example, adult limpets living at the highest tidal levels are exposed to the air for the longest periods and are more thermally stressed than those lower down the shore. Their shells tend to be paler, taller, and have more pronounced ridges than the shells of limpets living lower down the shore. For a limpet of a given volume, an individual with a taller shell has smaller contact with the substratum, thus reducing conduction. Relatively tall shells also project upwards where air flow is usually faster, facilitating convective cooling. Deep ridges on the shell also increase convective cooling.

Figure A Schematic of heat transfer between a limpet and its external environment.

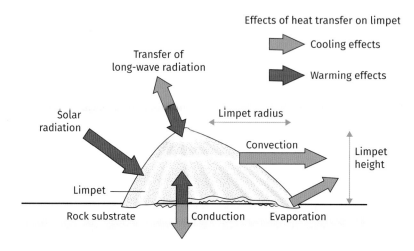

Adapted by Michael Kent from Denny, M. W. and Harley, D. G. (2006) Hot limpets: predicting body temperature in a conductance-mediated thermal system. The Journal of Experimental Biology 209:2409–2419. Published by The Company of Biologists 2006doi:10.1242/jeb.02257

Wind-tunnel experiments carried out by Mark Denny and Christopher Harley at Stanford University's Hopkins Marine Station using model limpets have shown that being relatively tall contributes more than 90 per cent of the convective cooling. Ridges have a significant cooling effect only at high wind speeds.

❓ Pause for thought

Water has a high latent heat of vaporization. This enables organisms to use evaporation as a cooling mechanism in hot environments. Limpets can sometimes be seen raising their shells when exposed to air on a hot day. Although this behaviour results in evaporative cooling, such behaviour is regarded as being of dubious significance as an adaptation to the thermal environment by intertidal biologists. Suggest reasons why, and consider why juvenile limpets are more vulnerable to thermal stress than adults.

Solvent properties and salinity

The dipolarity of water molecules makes seawater an excellent solvent. It readily dissolves any polar and ionic substances that enter it. Consequently, there's an amazing amount of salt in the sea. On average, each litre contains about 35 grams. That may not seem much, but if all the salt from the World Ocean was extracted and spread evenly over the Earth's surface, it would form a layer about 160 metres thick—more than enough to cover even the tallest trees.

Sodium chloride (table salt) accounts for about 78 per cent of the salts dissolved in seawater. But magnesium chloride, magnesium sulfate, calcium sulfate, and potassium chloride are also major components. This makes sea salt an incredible mixture of chemicals; it even includes the salts of precious metals such as silver, lithium, and gold. Most salt in the sea comes from the land when rocks and soil are worn away by rain. The rain runs into streams and rivers which ultimately make their way to the sea into which they deposit their salt. An estimated 4000 billion kilograms of salt are carried by rivers into the sea each year. The proportion of the different major salts is fairly constant throughout the oceans.

The saltiness of a body of water is referred to as salinity by marine biologists and oceanographers. Salinity is defined as the dissolved inorganic salt content of water measured in grams of salt per kilogram of water. It varies considerably over time and location. For example, a rock pool at the top of the shore, isolated by tides for days or weeks, can have a very high salt content (hypersaline) due to evaporation, and very low salt content (hyposaline) when heavily rained upon. As water evaporates from shallow rock pools, it leaves dissolved salts behind in the remaining water. Every rock pool, therefore, becomes saltier as its water evaporates and more dilute with the addition of rainwater. But shallow pools high on the shore are more prone to salinity changes than those lower down.

Although the average salinity of the world's oceans is 35 grams of salt per kilogram of seawater, the salinity of individual oceans and their component seas often differ. The Mediterranean Sea is generally saltier than the Atlantic Ocean. Arctic waters are more dilute than those on the equator. Surface salinity varies depending on loss of water by evaporation, and gains from precipitation, river inflow, and the melting of sea ice. The overall salinity of the global ocean remains relatively constant because fresh supplies of salts are continually being added to the oceans from rivers at roughly the same rate as various physical, chemical, and biological processes remove them.

Marine organisms living in intertidal rock pools or in estuaries have to be able to cope with variable salinities (see Case study 3.3).

Case study 3.3
Adaptations of shore crabs to osmotic stress

The Shore crab (*Carcinus maenas*; Figure A) has the reputation of being one of the toughest marine creatures in the world, except when it has to discard its body armour in order to grow. During moulting, its soft parts are exposed to the external medium and it becomes vulnerable to predators. For protection, it usually finds a secure place under rocks or in crevices in which to complete the moult and harden its new shell.

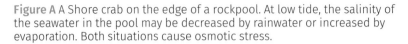

Figure A A Shore crab on the edge of a rockpool. At low tide, the salinity of the seawater in the pool may be decreased by rainwater or increased by evaporation. Both situations cause osmotic stress.

Photo: © Michael Kent

Shore crabs not only inhabit intertidal rock pools but they also penetrate into the brackish water of estuaries. They are able to cope with variable salinities in the external environment by tolerating changes in the concentration of solutes in their body fluids, and by a combination of behavioural, physiological, and structural adaptations.

Professor John Davenport, when at Bangor University, carried out some ingenious choice-chamber experiments that confirmed the ability of Shore crabs to detect differences in salinity and to move into a salinity that was less stressful. He concluded that when a rock pool becomes too saline or too dilute, the Shore crab is able to move into a pool that has less osmotic stress. One of the interesting aspects of the Shore crab's response to stressful salinities is that it carries out this behaviour even though it has the ability to maintain a relatively stable concentration of salts in its body fluids by physiological adaptations. In a hypersaline environment it is able to actively transport sodium ions across its gills and out of its body. *Carcinus maenas* also maintains a relatively constant body fluid concentration in dilute waters, enabling it to penetrate into the brackish waters of estuaries. Again, the gills play an important osmoregulatory role, but in a hypotonic environment they actively transport ions into the body.

The Shore crab is also able to change the permeability of its gills to water. In isotonic environments, it actively absorbs water. In a dilute, hypotonic environment, when osmoregulation begins, water absorption is suspended, probably due a fall in the passive permeability of the gills to water.

❓ Pause for thought

Suggest reasons why the Shore crab evolved a behavioural adaptation to salinity when it is able to tolerate wide ranges in external salinities, and has physiological and structural adaptations to keep the concentration of salts in its body fluids constant.

Dissolved substances

The most abundant gases dissolved in seawater are nitrogen, oxygen, and carbon dioxide. Their main source is the atmosphere. The gases enter or leave the ocean across the air–sea interface, and are transported within the ocean by physical processes. The ocean can act as either a source or a sink for atmospheric gases. The levels of dissolved gases vary with latitude, because gases are more soluble in cold water than in warm water.

Dissolved oxygen

Oxygen is not so freely available in the marine environment as in the terrestrial environment. Because atmospheric oxygen diffuses slowly through the surface of the sea, it is not rapidly replaced when it is used up in respiration and bacterial decomposition. Consequently, oxygen levels tend to decrease with depth until they reach a minimum level, after which the levels rise again until they reach a maximum in cool deep waters. In the tropics about twice as much oxygen is dissolved in cold deep waters than in warm surface waters. The solubility of oxygen and other gases also increases with pressure.

The oxygen minimum usually coincides with the thermocline (Figure 3.4) as this area of abrupt change in temperature acts as a barrier to the mixing of oxygen-rich lower layers with oxygen-depleted surface layers. Upwelling (see Figure 2.12) disrupts the thermocline and brings oxygen-rich waters towards the surface.

Although seawater at great depths is typically oxygen-rich, seawater percolating into soft muddy substrates is usually oxygen-depleted at any depth. Benthic organisms, such as polychaete worms, burrowing and crawling in mud and ooze, often have respiratory pigments that can extract oxygen from poorly aerated substrates. For example, the haemoglobin in the lugworm, *Arenicola marina*, has an oxygen dissociation curve well to the left of that of humans (Figure 3.5).

This molecular adaptation of the lugworm to its marine environment enables it to absorb oxygen with maximum efficiency in its oxygen-depleted habitat.

Figure 3.5 Oxygen dissociation curves for lugworm and adult human haemoglobin.

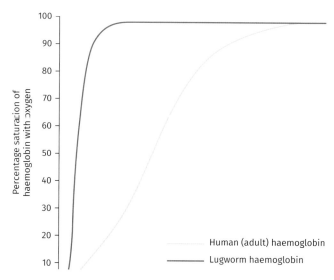

Dissolved carbon dioxide

Dissolved carbon dioxide plays an important part in maintaining the pH of the Ocean. The average pH of seawater is about 7.8 with a range from 7.5 to 8.5. Seawater is therefore slightly alkaline (or basic), whereas pure water has a neutral pH of 7.0. Carbon dioxide reacts with water to form carbonic acid which dissociates into hydrogen carbonate ions and carbonate ions to produce a system that resists changes in pH (see Figure 3.6)—in other words, it acts as a buffer. Although seawater acts as a buffer, the addition of acid does lower its pH, but at a much lower rate than if it were pure water.

Figure 3.6 The interactions between atmospheric carbon dioxide and the water of the ocean.

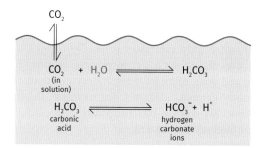

Dissolved Organic Matter

Dissolved Organic Matter (DOM) refers to carbon-containing substances in aqueous solution. There was a time, not so long ago, that DOM was barely mentioned in scientific accounts about factors affecting life in the sea. It was thought to be simply a wasted source of carbon and energy. Now it's recognized as a major component of food webs.

It's estimated that up to 25 per cent of primary production in the marine environment is not transferred directly to consumers, but is leaked into the water as DOM. Viruses are partly responsible for this leakage by infecting and killing their photosynthesizing hosts, causing them to disintegrate and release their body contents into the sea. The resulting DOM is consumed by bacteria which, in turn, are consumed by heterotrophic micro-plankton, such as flagellates and ciliates. This flow of energy between photosynthesizing picoplankton and nanoplankton (see Table 1.1) is called the microbial loop. It's a major feature of the Arctic food web where algae and other photosynthetic organisms in the ice leak DOM into brine channels (Figure 1.2).

When kelp and other macroalgae die they disintegrate and, collectively, release huge amounts of DOM and Particulate Organic Matter (POM) into the sea. The DOM and POM form a substrate on which bacteria can grow and feed. Sometimes this mix of DOM, POM, and bacteria is whisked by waves and wind into a dirty looking off-white foam. If the foam is blown inshore, it blankets sand and rocks and dries into a nutritious scum. This sea foam (Figure 3.7) is a natural phenomenon, not the result of pollution as is commonly misconceived.

Figure 3.7 Sea foam accumulating after an autumn storm on a shore in north Cornwall.

Photo: © Michael Kent

The bigger picture 3.1
What's in a name?

The pre-eminent way of communicating scientific information is through peer-reviewed publications. Typically, an article, known as a 'paper', is submitted to the editor of a journal who sends it out to several scientists who work in the same field as the paper's author (or authors). The reviewers scrutinize the article and inform the editor whether they think the paper is good enough to be published. Only articles that meet good scientific standards are accepted for publication.

The **peer-review** process is a tried-and-tested means of vetting a scientific paper before publication. After publication, papers are cited as sources in future publications. The number of times a paper is cited is one measure of its global impact on the scientific community.

Farooq Azam and Tom Fenchel were lead authors of a paper entitled *The ecological role of water-column microbes in the sea*. The paper has had an enormous impact, having been cited more than 5800 times up to June 2020. And it was awarded the John H. Martin Award for 'a high-impact paper in the aquatic sciences' by the American Society of Limnology and Oceanography. In 2008, 25 years after its publication, Tom Fenchel wrote a review that included an analysis of the high impact of the paper. He noted that the paper didn't contain anything that was particularly new: it mainly summarized previous work, as the general idea of the existence of a microbial loop (Figure A) had been expressed before.

Concerning the huge impact of the paper, Tom Fenchel reported that the real reason for the impact was that he and his co-authors gave the phenomenon a descriptive name. In just a couple of words, 'microbial loop' described that plankton food chains are more complex than previously understood, and that microbes play a much greater role in food chains than had been acknowledged in the textbooks of the day. He concluded 'The term "microbial loop" suddenly clarified what we and others were working with and it comes as a label for studies of marine and freshwater environmental microbiology, studies on tiny protozoa, and much more. So a name does matter after all!'

Figure A The microbial loop—DOM refers to dissolved organic matter (also called DOC, dissolved organic carbon).

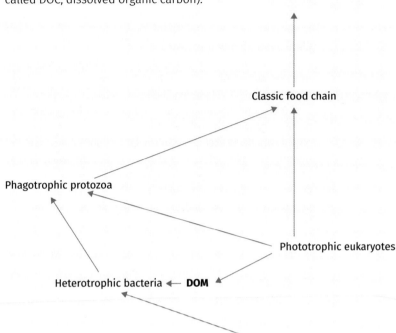

Source: Azam, F., Fenchel, T., Field, J. G., Gray, J. S., Meyer-Reil, L. A., Thingstad, F., The Ecological Role of Water-Column Microbes in the Sea. Marine Ecology Progress Series 10:257–263

❓ Pause for thought

Think of another biological phenomenon that has been given a name that has captured the imagination.

One criterion that a paper must reach before being published is that it acknowledges and builds upon the work of others in the field. What other criteria do you think are important?

Being published does not mean the contents of a paper are correct or conclusive. Suggest how a paper that contains incorrect contents might be cited many times.

Dissolved nutrients

Nitrogen and phosphorus are essential elements for the growth of primary producers. Nitrogen is absorbed by marine photosynthesizing organisms directly from the surrounding seawater in the form of ammonium, nitrate, and nitrite ions. Nitrogen-fixing bacteria are the primary source of these ions which they obtain by 'fixing' nitrogen gas dissolved in seawater. The nitrogen-containing ions are then released when the bacteria die.

Phosphorus leaks into the marine environment naturally, much of it in the form of phosphate ions which can be readily absorbed by photosynthesizing organisms.

Once nitrogen- and phosphorus-containing compounds are assimilated into the bodies of photosynthesizing organisms, they become available to higher trophic levels in marine food webs.

Nitrates and phosphates are abundant in deep-ocean waters due to dead bacteria continually sinking downwards. When the euphotic zone is well-mixed with the deeper layers, nitrogen and phosphorus nutrients are not usually limiting factors to photosynthesis. However, the availability of these nutrients declines when the euphotic zone becomes separated from the nutrient-rich deep ocean by a thermocline (see Figure 3.4). In such circumstances, nitrogen and/or phosphorus can become limiting factors for photosynthesis. Nitrogen is often the limiting factor in oceanic environments and phosphorus in coastal waters.

Phytoplankton also require iron to carry out photosynthesis. Iron is an element essential for the production of chlorophyll. In the marine environment, it comes mainly from iron-rich dust that is blown into oceans by desert dust storms. In most parts of the global ocean, iron is not normally a limiting factor. However, in a few parts of the open ocean photosynthesis does not take place despite nitrogen and phosphorus nutrients being readily available. Iron is the limiting factor, and, as iron deficiency prevents the production of chlorophyll, these parts of the ocean are called High Nutrient–Low Chlorophyll (HNLC) regions.

Density and viscosity

The physical constraints that determine the size of organisms on land and in the sea are very different. The density of seawater is much higher than that of air. The relatively high density of seawater allows huge animals like the Blue whale and Giant squid to thrive in the marine environment. It also allows seaweeds as large as the Giant kelp to extend tens of metres upwards from the seabed to the surface without any special means of support, apart from air bladders.

A decrease in temperature or an increase in salinity increases the density of seawater, causing it to sink below warmer, less saline water. High pressure also increases the density of seawater, but only slightly compared to the effects of temperature and salinity.

When seawater freezes, its water molecules move so slowly that they become locked together by hydrogen bonds to form solid crystals. Water molecules in ice crystals are further apart than in liquid water, explaining why water expands when it freezes and becomes less dense as a solid than as a liquid. Consequently, when water freezes over an extensive area, a layer of floating ice forms on top of liquid seawater and insulates it so that it doesn't continue to freeze.

Viscosity is a measure of a fluid's resistance to flow. It is ultimately a result of the intermolecular forces between the fluid's molecules. Due to hydrogen bonding between its molecules, water's viscosity is unusually high compared to other liquids with similarly sized molecules.

Seawater viscosity is slightly greater than that of pure water. It increases gradually with a rise in salinity, and to a much greater extent with a fall in temperature. The relationship between viscosity and pressure is complicated and is affected by temperature. Usually viscosity increases with pressure, but water behaves anomalously below about 30 °C when, at low pressures, increasing pressure reduces viscosity rather than increasing it. Water viscosity passes through a pressure minimum and then increases. Changes in viscosity can have a significant effect on the movement of organisms through seawater.

Size matters

The physical constraints that determine the size of organisms on land are quite different than those that determine size in marine organisms. The density of seawater being so much greater than that of air provides the support which allows huge animals and algae to live in the sea.

Since water is much more viscous than air, movement is quite different in the two fluids. For an object moving through a fluid, inertia tends to keep it moving whereas viscosity tends to stop it. Basically, the higher the viscosity of a fluid, the more slowly an object will move through it. However, movement through a liquid also depends on the size, shape, and velocity of the objects. These factors are taken into consideration in a quantity called the Reynolds number (*Re*).

Re is a ratio of the inertial forces on an object to the viscous forces acting on it. Inertial forces depend on the density of the fluid, and the velocity, size, and shape of the object. When calculating *Re*, viscosity is represented as kinematic viscosity (calculated as density divided by viscosity), measured in the same units as velocity. By using kinematic viscosity, *Re* can be expressed as a dimensionless quantity and calculated as:

Re = (particle velocity × particle size) / kinematic viscosity

Generally, the smaller the organism, the smaller is the *Re*: a Blue whale swimming at 10 ms^{-1} has an *Re* of about 300 000 000; a large fish swimming at the same speed has an *Re* of about 30 000 000; an adult person swimming at 0.3 ms^{-1} has an *Re* of about 54 000; a bacterium swimming at 0.01 ms^{-1} has an *Re* of about 0.00001.

In large pelagic organisms, size and speed are usually positively correlated. These organisms typically have high *Re* values (much more than 1) that indicate turbulent flow of water over their bodies. Inertial forces and gravity have the greatest effect on their movement.

For small marine organisms that have a very low *Re* (much less than 1), viscous forces are dominant. For these organisms, seawater acts as a highly viscous fluid. Moving through it produces little momentum. It has been likened to someone swimming in a pool full of treacle and having to swim at less than 1 cm per minute in a circular pattern. Under these conditions the swimmer would move about 100 metres in a week.

Stopping distances for organisms with a large *Re* are quite different to those for organisms with a small *Re*. If a fast-moving large fish were to suddenly stop swimming, it would take several metres for it to come to a halt, but a swimming bacterium would come to a halt in a distance less than the diameter of a hydrogen atom.

Re also affects how long an inert organism can remain at a particular depth in the water column. Gravity acts on every object, but in seawater it

Figure 3.8 Zooplankton and phytoplankton of various sizes and shapes.

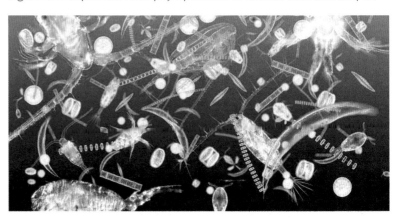

Photo: © Richard Kirby

has a greater effect on organisms with a high *Re* than those with a low *Re*. Consequently, a dead whale will sink to the seafloor relatively faster than a member of the microplankton.

For phytoplankton (see Figure 3.8), it's a great advantage to stay in the euphotic zone as long as possible so that they can maximize their harvesting of light for photosynthesis. Therefore, it's not surprising that they have evolved various morphological adaptations to increase their *Re*. Being small is one obvious adaptation. However, it has its disadvantages: the most important is that being small limits the amount of chlorophyll an individual member of the phytoplankton can contain. Intraspecific differences in size are common in phytoplankton, with smaller individuals living in areas where solar radiation is optimal, and larger individuals living in areas where it is suboptimal.

One peculiar aspect of the rate of sinking of phytoplankton in seawater is that dead algae sink up to five times faster than live, immobile cells despite dead and live cells appearing to the naked eye to be the same in shape, size, and structure. Differences in sinking rate were attributed to some mysterious 'vital factor' until detailed microscopical studies revealed that living algae can produce peculiar organic or chitinous bristles protruding from the cell margins; even under the light microscope, these bristles are barely visible.

Compressibility

Liquid seawater, like pure water, is difficult to compress. This incompressibility has two major consequences:

- The density of sea water changes very little however deep you go, because the liquid is hardly compressed. A mile down, the volume of seawater is only 1 per cent less than it is at the surface! As a result of the incompressibility of water, even in the deepest parts of the ocean the density of the seawater is only a little higher than that at the surface, and there is only a 1.8 per cent decrease in volume.

- Because water—pure or sea water—is difficult to compress, it can be used to produce movement in closed, fluid-filled systems. This property forms the basis of the hydroskeletons in many marine organisms.

Pressure

Water is a dense medium and exerts a significant pressure on any object submerged in it. Pressure is defined as 'force per unit area'. It is commonly measured in bars. One bar is equal to 100 kPa—roughly equivalent to one atmospheric pressure at sea level. The pressure in the ocean increases approximately by one bar for every 10 metres increase in depth. An organism living at a depth of 100 metres on the continental shelf will experience pressures about ten times greater than an organism living in the intertidal zone. Organisms living in oceanic trenches, the deepest parts of the ocean, experience a pressure more than a thousand times greater than that experienced by shore dwellers.

At very high pressure, protein molecules are compressed, denatured, and altered in shape, structure, and chemical activity. Animals in the depths of the abyss have evolved molecular adaptations that enable the proteins to function at high pressure. For example, deep-sea fish have a different form of actin in their muscles than relatives living higher in the water column.

Very high pressures also make cell membranes more rigid. Deep-sea organisms compensate for this by having more unsaturated lipids in their membranes.

Increasing pressure with water depth has its greatest effect on the volume changes of gases contained within marine organisms, especially air-breathing vertebrates that dive. Diving marine reptiles, birds, and mammals all have structural, physiological, and behavioural adaptations that enable them to dive deep. We are going to look at the adaptations of just one, Cuvier's beaked whale, which has the deepest recorded dive of any air-breathing vertebrate at 2992 metres (see Figure 3.9).

Cuvier's beaked whale spends much of its life in deep water hunting for prey, mainly squid and deep-sea fish. A single dive may last more than two hours (the record dive was timed at 137 minutes). During this time, the whale has to survive on a single breath of air.

When diving deep, any air-breathing vertebrate has to overcome two major challenges: firstly, it has to be able to deal with extreme changes in pressure; secondly, it has to be able to store enough oxygen to hunt successfully.

Increasing pressure shrinks the vertebrate lungs until at depths of 200 metres they will have collapsed completely. On reaching this depth, the pressure compresses any air in the alveoli, increasing the amount of gases dissolved in blood and other tissues. This is not a problem until the mammal rises to the surface. Then, decreasing pressure causes gases in the blood to come out of solution, releasing nitrogen bubbles into the blood stream. In human divers breathing from a Self-Contained Underwater Breathing Apparatus (SCUBA), a rapid ascent causes bubbles of nitrogen gas to escape

Figure 3.9 Cuvier's beaked whale (*Ziphius cavirostris*): the deepest diving marine mammal.

Photo: HeitiPaves/iStock

into the blood stream; this can lead to the 'bends', a painful blockage of blood vessels that can be fatal.

Scientists are not completely sure how Cuvier's beaked whale copes with this problem. One suggestion is that before diving it collapses its lungs in such a way that the air is forced out of the alveoli and stored in the nasal cavities and windpipe where nitrogen cannot be absorbed into tissues. By whatever means Cuvier's beaked whale avoids the 'bends', it does not explain how it survives without air for so long.

Cuvier's beaked whale, in common with other cetaceans, has a body specially adapted to store oxygen in blood and tissues rather than carrying it in their lungs as we do. The whale has extraordinarily high levels of haemoglobin in the blood. It also has high levels of myoglobin, a protein that stores oxygen inside muscle. The very high levels of haemoglobin and myoglobin make the blood and muscles appear almost black in colour.

All mammals (including humans) have a reflex reduction of heart rate (called bradycardia) when submerged, and they shunt their blood from the muscle and skin to the more vital organs, especially the heart and brain. But Cuvier's beaked whale and other cetaceans take this circulatory adaptation to its extreme limits, dramatically reducing the body's demand for oxygen.

As well as these extraordinary physiological adaptations, Cuvier's beaked whale has structural adaptations to diving deep, including indentations in the body surface that act as pockets for their flippers, which enable the whale to be shaped like a torpedo. Streamlining its body shape helps the whale to move with maximum efficiency through water and use its oxygen stores for as long as possible.

Electrical conductance

Pure water has no free ions. Consequently, it is an excellent insulator and does not conduct electricity. The addition of dissolved salts enables electrically charged ions to flow freely. The saltiness of seawater makes it a good conductor of electricity, a property used in electric fish such as *Torpedo torpedo* (Figure 3.10) to detect and capture prey.

Torpedo rays have special electric organs (two large, kidney-shaped organs located beneath the skin on either side of the head) derived from modified nerves and muscles. These organs can emit electric charges that are sufficiently powerful to stun and even kill prey.

Rays, as well as all other elasmobranchs, are able to detect electricity through special organs, called Ampullae of Lorenzi, located over the top and sides of the head. All marine animals emit electrical charges when moving, therefore these electro-sensory organs enable elasmobranchs to detect their prey even in dark environments where they can't be seen. As well as being used to stun prey, fish electric organs are also used for defence, communication, and navigation.

Figure 3.10 The Torpedo ray, *Torpedo torpedo*—just one of a number of marine species which rely on the conductivity of sea water to control and capture their prey.

Photo by Roberto Pillon

Sound transmission

One result of water's low compressibility is that sound waves travel through seawater with four times the velocity and less absorption of energy than through air. Marine animals take advantage of the high quality of sound transmission in the sea by producing various types of noise to communicate with one another, to attract mates, and to deter enemies.

Pistol shrimps (*Alpheus* spp.) are among the loudest creatures in the marine environment. One of the pair of claws in a pistol shrimp is massive. It can be snapped shut so quickly that it creates a gas bubble which implodes instantaneously. The sudden implosion produces a flash of light and a sonic shockwave through the water which can stun and kill prey. The sound associated with the shockwave exceeds 200 decibels, far louder than a gunshot.

Even louder than pistol shrimps are sperm whales (*Physeter macrocephalus*; see Case study 6.3, Figure A). By forcing air from sacs located in the head just below the blowhole, they produce clicking noises that can reach 230 decibels. Cetaceans have excellent hearing. They use sound for communication, echolocation, and prey location.

Transparency to visible light

Seawater's transparency to visible light is vitally important to photosynthesizing marine organisms. The amount of sunlight reaching the surface of the sea varies with weather, time of day, season, and latitude. Most of the light striking the sea warms the surface water or is reflected. A small amount penetrates the surface. Light intensity decreases with depth until no light penetrates at all.

On a sunny day, the depth to which light penetrates seawater depends largely on the amount of plankton and suspended matter it contains. In murky coastal waters, sunlight may penetrate only a few tens of metres. Even in clear oceanic waters, there is little sunlight below a few hundred metres (see Figure 3.11).

Figure 3.11 Transmission of visible light in the sea showing the depth at which different colours of light penetrate ocean waters.

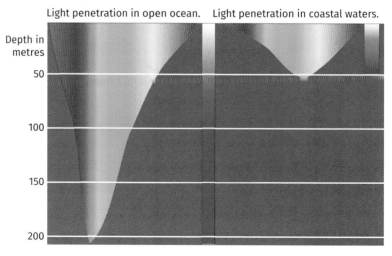

Image courtesy of Kyle Carothers, NOAA-OE

Not only does the quantity of light diminish with water depth, its spectral composition also changes. In the open ocean, the red band becomes extinct within the first few metres, followed in turn by the orange, yellow, and green bands. The blue band penetrates the deepest.

When I asked a Professor of Physical Oceanography why the blue band penetrates the deepest, he said there's no easy answer, but basically it results from there being two main mechanisms in water that absorb light. The first mechanism involves the vibration of water molecule bonds (mostly the stretching of O–H bonds). It is the primary cause of absorption from the infrared to blue parts of the spectrum, it has its greatest effect on infrared and red light, and becomes less effective with decreasing wavelength. The second mechanism involves electrons absorbing energy and causing them to move further away from nuclei (a process called electron transition). As wavelength decreases beyond blue light, absorption by electron transition begins to become more important, causing a rise in absorption from the violet and into the ultraviolet ranges.

The net result of these two mechanisms is that light absorption is greatest towards both ends of the visible spectrum, and least in the middle part of the spectrum—hence the deepest penetration of blue light.

Photosynthesizing organisms have evolved many adaptations to optimize their position in the photic zone. They often have accessory photosynthetic pigments to improve photosynthetic efficiency in areas of low light intensity or where the spectral composition is not ideal for the functioning of chlorophyll, the primary photosynthetic pigment.

Marine scientists use a wide range of instruments to study light in the sea. The instruments range from simple Secchi discs to submersible light meters and sophisticated spectral radiometers carried by satellites.

The Secchi disc measures the transparency of seawater. Typically, a plain white, circular disc 30 cm in diameter is lowered into the sea and the depth at which it disappears, called the Secchi depth, is recorded. This depth is inversely proportional to the transparency of the water.

Measurements in the field have shown that Secchi depth equals about one third of the depth of the euphotic zone. The greatest published Secchi depth is 80 metres. It was measured in the Weddell Sea at the end of an Antarctic winter when there were no phytoplankton in the water. The Secchi depth indicates that the euphotic zone in the Weddell Sea was about 240 metres deep when the measurement was made.

Submersible light meters quantify the spectral composition of the light field at different depths in the ocean and measure how light is absorbed and scattered in water over short distances. Spectral radiometers on Earth-orbiting satellites measure the colour of light reflected from the surface layer of the ocean. These colours literally reflect the distribution of phytoplankton (Figure 3.12). By recording spectral changes of the ocean surface, scientists can monitor how phytoplankton populations change over time.

Figure 3.12 Satellite image showing a milky white phytoplankton bloom in the English Channel off the southwestern tip of England. The milky turquoise colour is due to sunlight reflecting from billions of the coccolithophore *Emiliania huxleyi*, the surface of which is composed of white calcium carbonate scales.

Source: Copernicus Sentinel 2 data processed by the NERC Earth Observation Data Acquisition and Analysis Service (NEODAAS), Plymouth Marine Laboratory.

Chapter summary

- There are two main theories about the origins of seawater—that the Earth might have captured water from asteroids and comets which collided with it or that the water was released explosively in huge quantities from inside the Earth itself, during a period of intense volcanic activity.
- Two basic properties that make seawater special are dipolarity and hydrogen bonding.
- Other special properties that make seawater special include its high surface tension; its thermal properties, including its high specific heat capacity, latent heat of vaporization, and its freezing properties; its solvent properties; its relatively high density and viscosity compared to pure water; its high compressibility; the increase in its pressure with depth; and its ability to conduct electricity, and transmit sound waves and solar radiation.
- Adaptations of marine organisms to living in seawater are discussed. These include adaptations to the thermal environment, to salinity variations, to moving and floating in water, to the photic environment, and to diving from the surface into deep waters.

Further reading

Andersen, N.M., and Cheng, L. (2004) The marine insect *Halobates* (Heteroptera, Gerridae): biology, adaptations, distribution and phylogeny. *Oceanography and Marine Biology Annual Review.* CRC Press.

Branch, G., and Branch, M. (2018) *Living shores: interacting with southern Africa's marine ecosystems.* Struik Nature.
An authoritative and fascinating textbook that brings to life the rich biodiversity of the shores around South Africa.

Cox, B., and Cohen, A. (2013) *The Wonders of Life.* Harper Collins.
A beautifully illustrated book by Professor Brian Cox and Dr Andrew Cohen that uncovers some mind-boggling secrets of life in the sea and throughout the cosmos.

Kirby, R. (2012) *Ocean drifters: A secret world beneath the waves.* Firefly.
Stunning close-up photographs of the secret world of plankton.

Discussion questions

3.1 Suggest why there are no truly marine amphibians.

3.2 How might the size of microscopic, living, motile organisms be measured?

3.3 We have considered how size and shape affect the movement of plankton in seawater. How might these parameters affect other processes such as metabolic rate, nutrition, photosynthetic efficiency, and predator–prey relationships?

4 WHAT IS MARINE BIODIVERSITY?

The term 'biodiversity' was first coined by Dr Walter G. Rosen, sometime in 1985. He invented the term as a convenient contraction of 'biological diversity' when planning for the 'The National Forum on BioDiversity'.

Sixty leading biologists, economists, and experts from many other fields attended the Forum. It coincided with an upsurge of interest among scientists and non-scientists in matters relating to the diversity and conservation of wildlife. An estimated audience of 5000 to 10 000 took part in the teleconferenced discussions held on the final evening of the Forum.

The proceedings of the Forum formed the basis of an influential book entitled *Biodiversity* edited by E.O. Wilson and Frances Peter (see Further reading). *Biodiversity* became a best seller and the book is still cited today. It had an enormous impact on the public's understanding of the diversity of life, the economic value of biological resources, and the need for international conservation. The success and popularity of the book is said to have launched the word 'biodiversity' into general use. Today, biodiversity is one of the most commonly used terms in the life sciences— it has become a household word. Biodiversity has also become a scientific discipline in its own right, with its own university courses and dedicated scientific journals.

Although mainly a terrestrial biologist, E.O. Wilson (see Figure 4.1) is an advocate for the biodiversity of the whole planet. When interviewed on his 90th birthday, he put forward fresh arguments for setting aside half of the Earth's terrestrial and marine environments for other species. He made this audacious biological conservation proposal in his 2016 book *Half-Earth*.

Figure 4.1 E.O. Wilson, the world's leading expert on ants, is regarded by many as the 'father of biodiversity'.

Photo by Jim Harrison - PLoS, CC BY 2.5, https://commons.wikimedia.org/w/index.php?curid=4146822

Biodiversity defined

Search and you will find hundreds of definitions of biodiversity and its synonym biological diversity. But probably the most important is in the Convention on Biological Diversity (CBD). The CBD defines biodiversity as '... *the variability among living organisms from all sources including, inter alia ["among other things"], terrestrial, marine and other aquatic ecosystems and the ecological complexes of which they are part; this includes diversity within species.*'

The CBD was put together during the 1992 Earth Summit as part of an international Treaty. The Summit was held in Rio de Janeiro in 1992. It was convened by the United Nations General Assembly to increase understanding and raise awareness of biodiversity issues. Subsequently, the Treaty was signed by 150 nations.

The CBD definition of biodiversity contained within the Treaty has become the most important, widely used, and far-reaching in the world. It is usually cited in international legislation on the conservation of biodiversity.

Many other definitions of biodiversity are consistent with the CBD definition. For example, E.O. Wilson defines biodiversity as '... *all hereditarily based variation at all levels of organization, from the genes within a single local population or species, to the species composing all or part of a local community, and finally to the communities themselves that compose the living parts of the multifarious ecosystems of the world.*' But no single definition of biodiversity, not even that of the 'father of biodiversity' has gained universal acceptance.

Wilson's definition has been criticized as being so vague that it has led biodiversity to mean almost anything to do with the variety of life. Vagueness is not the only criticism levelled at biodiversity. It is also accused of being a heavily value-laden term.

A value-laden term?

As a term defined within a Treaty written to bind signatories to conserve wildlife and to reduce habitat loss, biodiversity has become inextricably linked to the objectives of the Treaty. These objectives are given in Article 1 of the Treaty as '... *the conservation of biological diversity, the sustainable use of its components and the fair and equitable sharing of the benefits arising out of the utilization of genetic resources, including by appropriate access to genetic resources and by appropriate transfer of relevant technologies, and by appropriate funding.*'

Currently, biodiversity is a term used widely by different groups of people in different ways. As John Spicer (Professor of Marine Biology at Plymouth University) wrote in the first edition of his book *Biodiversity*, '*It's a word frequently found on the lips of politicians, ecowarriors, broadcasters, business people, university students, your friends and acquaintances down at the pub or cafe, conservationists, and even school children. And yet trying to pin down exactly what all these different types of people mean by biodiversity is difficult. It seems to mean different things to different people.*'

If biodiversity is used in a scientific context, such as when referring to the variety of life in the marine environment, it should be made clear how it is measured. This is a topic to which we will return in Chapter 5. However, in most situations it's sufficient to know that biodiversity refers to 'the variety of life—in all its different forms and relationships'. However, even this definition is not as simple as it seems. For example, what is the 'life' to which it refers? Does it include viruses? (See The bigger picture 4.1 *Viruses and the meaning of life*.)

In the following sections we will be exploring how biodiversity, and marine biodiversity in particular, can be assessed at three main biological levels of structure: within species (genetic diversity), between species (species diversity), and between ecosystems (ecosystem diversity). Then we will look at functional biodiversity, in which biodiversity is viewed in relation to the things that organisms do and the role they have in communities or ecosystems.

The bigger picture 4.1
Viruses and the meaning of life

Viruses are probably the most numerous biological entities in the marine environment (see Scientific approach 1.1). Their importance is beyond doubt. They control the abundance of bacteria in the sea; they have a significant effect on geochemical and nutrient cycles, and food chains (see Figure B); and they can alter the course of evolution by transferring genes from one organism to another. However, even biologists, whose job it is to study life, cannot agree whether viruses are living or non-living. If living, they should be included in comprehensive counts of biodiversity; if non-living they should be excluded.

The question of whether or not marine viruses are part of marine life hinges on the meaning of the term 'life'. Although there is no universally accepted definition, scientists generally agree that most forms of life share certain fundamental characteristics: they are cellular in structure and able to feed, grow, excrete, respire, move, respond to external stimuli, and evolve. They are also, to some extent, able to control their own internal environment (a property called **homeostasis**).

Many biologists deem **autopoiesis**, the ability to reproduce and maintain itself, as the defining property of life. A complete virus particle (called a **virion**) typically consists of one or two molecules of DNA or RNA which form the core, enclosed in a protein coat called the capsid (see Figure A). This coat protects the viral genome and helps transfer DNA or RNA into the host. Some viruses have an outer layer or envelope surrounding the capsid. The envelope can be a complex mix of carbohydrates, lipids, and different proteins. It can even have a mass of surface fibres—see Figure A.

In their book *What is life?* Lynn Margulis and Dorion Sagan argue that only cells and things made of cells, from multicellular organisms to biospheres, are autopoietic and can metabolize. They recognize DNA molecules as being unquestionably important for life on Earth, but emphasize that the molecule itself is not alive. On the subject as to whether or not viruses are alive, they conclude that as viruses are not autopoietic, they are not alive.

Margulis and Sagan, however, came to their conclusion at a time when all known viruses were smaller than cells, and lacked sufficient genes and proteins to maintain themselves. They described the reproduction of viruses as being similar to that of digital viruses in computers, and regarded viruses as, at best, 'chemical zombies'. However, since Margulis and Sagan came to their conclusions, giant viruses large enough to be seen with a light microscope have been discovered.

Giant viruses, called Mimiviruses (see Figure A), containing nearly three thousand genes, have been found in the deep sea. Some of the genes encode for proteins that play an important role in metabolic processes in marine organisms. In a marine microbial food web (see Figure B), consumption by

Figure A i) A thin section electron micrograph of a marine Mimivirus with two virophages indicated by arrows. ii) Diagram to show the structure of a Mimivirus. The 'Stargate' is the position of the portal through which the Mimivirus releases its DNA. When open, the portal is shaped like a starfish.

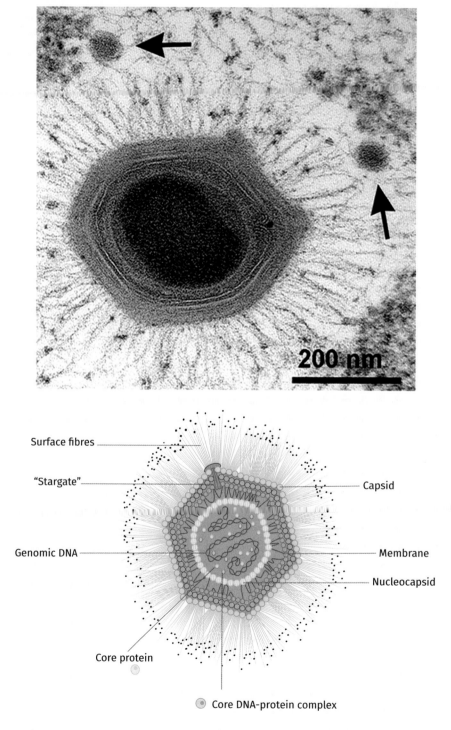

Source: Image i) made available courtesy of J.Y. Bou Khalil and B. La Scola, IHU Mediterranée Infection, France

Figure B A marine microbial food web, with the viral shunt highlighted.

grazers transfers carbon up the traditional food chain into higher trophic levels. By contrast, the viral shunt produces dissolved organic matter that can be consumed by other prokaryotes, essentially serving as a marine microbial recycling system that stimulates nutrient and energy cycling. Although Mimiviruses still require a host cell in which to replicate, and their metabolism is limited, their genetic complexity is said to 'blur' the distinction between living and non-living entities.

Perhaps Mimiviruses should be regarded as the ultimate parasites.

❓ Pause for thought

Suggest why conforming to a list of characteristics is not a completely satisfactory way of defining life.

It has been argued that the only satisfactory definition of life lies in the ability to evolve independently. A 2020 article in *Science*, a magazine for the American Association for the Advancement of Science (AAAS), reports that scientists have produced software based on the Darwinian principle of 'survival of the fittest', so that the Artificial Intelligence (AI) programs incorporated in robots are able to improve from one generation to the next, without human interference. It is theoretically possible to program robots to build new robots with the evolving programs. Should self-replicating robots with an evolving AI be regarded as living and therefore part of biodiversity? If such a robot were considered to be living, what would be the status of a computer virus that is implanted in it?

Genetic diversity

Genetic diversity is defined as the variation in the genetic information within and among individuals of a population, species, or any other level of biological organization.

The genetic diversity of the global ocean has been called the ocean genome. It's the sum of all the information in the genes of all the individuals of all the species that make up marine life. Its economic importance is reflected in a report presented to the High Level Panel for a Sustainable Economy (HLP). The HLP was established by 14 heads of government to seek ways of sustaining the health and wealth of the global ocean, and building a better future for all people and the planet.

The report to the HPL defined the ocean genome as '... *the genetic material present in all marine biodiversity, including both the physical genes and the information they encode'*.

One of the major scientific justifications for studying genetic diversity within an ecosystem is to test the assumption, implicit in the report to the HPL, that the higher the genetic diversity, the better a natural population is able to adapt to changing environmental conditions such as climate change. Case study 4.1 discusses how this assumption has been tested in seagrasses.

Case study 4.1
Genetic diversity and resilience in seagrasses

Seagrasses are true plants. Just like terrestrial grasses, they are monocotyledons and, again like other grasses, they have leaves, roots with veins, and reproduce using flowers and seeds.

Seagrass species (of which there are more than seventy) are also known as eelgrass, turtle grass, tape grass, shoal grass, and spoon grass, depending on the shape of the species and where it is found.

In 2016, the entire genome of one species, *Zostera marina* (commonly called eelgrass), was sequenced. Knowing the genome provides genetic information that enables research scientists to understand how these marine flowering plants are genetically adapted to environmental stresses, such as rises in sea temperature brought about by climate change.

Zostera marina is the most abundant eelgrass species in the northern hemisphere. It plays a critical structural and functional role in many coastal ecosystems. It is an important ecosystem engineer. It provides a nursery environment for a large variety of shallow water species, including seahorses (Figure A) and cuttlefish. It can form dense meadows from 1 to 5 metres in depth; some meadows are so extensive they are visible from space. Eelgrass meadows help protect against coastal erosion and they increase water clarity by reducing wave energy, trapping particles, and stabilizing sediments. They also play an important role in recycling nutrients, and they help reduce the impact of climate change by sequestering carbon dioxide from the atmosphere.

Anneli Ehlers and her co-researchers investigated the effect of genetic diversity in the eelgrass on its resilience to climate change (see Further reading).

Figure A A seahorse, just one of the many species that live in seagrass beds.

Figure B The effect of sea temperature and genetic diversity in eelgrass. A rise in sea temperature tends to cause a reduction in the growth of eelgrass shoots, resulting in negative effects on ecosystem functioning.

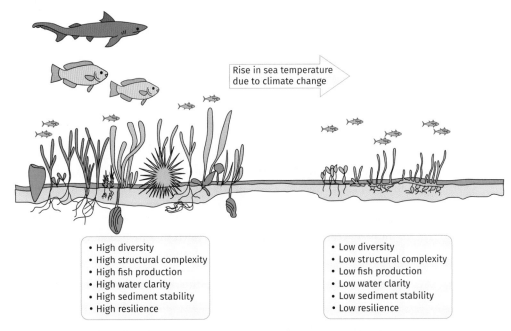

Rise in sea temperature due to climate change

- High diversity
- High structural complexity
- High fish production
- High water clarity
- High sediment stability
- High resilience

- Low diversity
- Low structural complexity
- Low fish production
- Low water clarity
- Low sediment stability
- Low resilience

They tested whether summer heat waves had negative effects on *Zostera marina*, and whether high genetic diversity may provide resilience in response to climatic change. In controlled experiments using marine mesocosms the researchers subjected eelgrass patches to an increase in seawater temperature over a 5-month period. A marine mesocosm is an artificial outdoor experimental aquatic system that simulates a set of conditions in a natural marine environment.

Ehlers and her co-researchers found a strong negative effect of warming on eelgrass shoot densities. They also found that an increase in genetic diversity had a positive effect—in other words, high genetic diversity made eelgrass more resilient to temperature change, and tended to moderate the effects of a temperature rise.

The results suggest that an eelgrass ecosystem will be negatively affected by the increases in summer temperature predicted by climate change models (see Figure B). They concluded that a higher genetic diversity within a population of *Zostera marina* might help to maintain the composition and functioning of the ecosystem, and provide resilience to environmental changes, such as increases in sea temperature.

❓ Pause for thought

Zostera marina and other seagrasses are the only fully marine flowering plants. Marsh plants and mangroves are semiaquatic and not fully marine. Why do you think so few flowering plants have colonized the marine environment?

Species diversity

Species is the most common level of biological organization at which marine biodiversity is assessed; it is the fundamental taxonomic unit in biological classifications. But before we consider the diversity of species, we need to tackle the thorny problem of deciding exactly what a species is.

Biological Species Concept

It is commonly understood that individuals belonging to the same species have similar morphological features and that they can interbreed to produce fertile offspring. This way of understanding species is known as the Biological Species Concept. Its major weakness is that it relies on individuals reproducing sexually. However, there are countless organisms that do not.

In practice, organisms are often allocated to a particular species on the basis of sharing similar morphological features, or are separated into two species because they look different. But this can result in organisms such as barnacles, which have life stages with different morphologies (Figure 4.2), being classified as separate species.

Figure 4.2 Acorn barnacle life cycle.

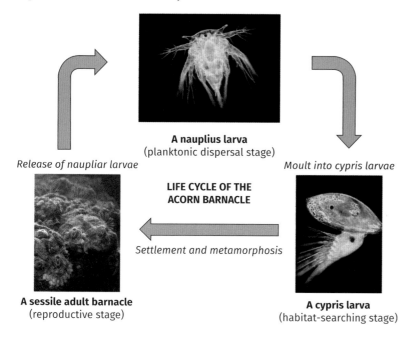

A nauplius larva
(planktonic dispersal stage)

Release of naupliar larvae   *Moult into cypris larvae*

**LIFE CYCLE OF THE
ACORN BARNACLE**

Settlement and metamorphosis

A sessile adult barnacle
(reproductive stage)

A cypris larva
(habitat-searching stage)

Images © Wim van Egmond

Adult and larval acorn barnacles (the nauplius and cypris stages) were regarded not only as belonging to separate species, but also to separate phyla. Adult barnacles were once classified as molluscs on the basis of their being encased in hard shells.

In 1830, Vaughan Thompson revealed the true crustacean nature of barnacles by tracing their life history. Adult acorn barnacles are sessile, filter-feeding, benthic organisms that live on rocky shores. The nauplii larvae are motile, temporary members of the plankton. Cypris larvae are the barnacle stage that settles on rocks. A cypris larva changes into an adult by a dramatic metamorphosis that results in its eyes and antennae being lost, and a cement gland being formed. This enables the cypris to glue itself to a rock or some other hard substrate.

There have been many attempts to define 'species', but no general agreement as to which definition is best. This supports Charles Darwin's contention that the term 'species' is an indefinable human construct '… *given for the sake of convenience to a set of individuals closely resembling each other'.*

The lack of an agreed definition has led some biologists to lump together organisms with slight variations into one species, whereas others tend to split them into separate species. This can cause great confusion in scientific communications and biological legislation, resulting in an organism appearing under more than one scientific name. However, DNA technology is changing our ideas of what a species is, and is helping taxonomists to define more precisely what it means to be a species.

Phylogenetic Species Concept

The discovery of the DNA double helix gave rise to species definitions that depended on small genetic differences between organisms. This has led to the Phylogenetic Species Concept in which species are distinguished on the basis of differences as small as 2% in their DNA.

DNA technology, involving nuclear and mitochondrial DNA sequencing, has also enabled the identification of cryptic species.

Cryptic species

According to the *Oxford Dictionary of Ecology*, the term 'cryptic species' refers to species which are apparently identical phenotypically (often to the point where individuals of such species are themselves unable to make the distinction) but that are incapable of producing hybrid offspring. For example, Moon Jellyfish are a common planktonic species in coastal waters around the world—they are found between latitudes of about 50 °N and 55 °S. Perhaps because of their global distribution, Moon Jellyfish have become popular research organisms for studies ranging from protein chemistry to ecology, and they are also among the most commonly displayed jellyfish in public aquaria. They are economically important because they feed on the young of many commercial species, and because they can block trawl nets and the water-cooling systems of power plants. Until the arrival of DNA sequencing, scientists thought all Moon Jellyfish belonged to one species. DNA analysis has shown that there are seven cryptic species, the chief of which are *Aurelia aurita, A. limbata* and *A. labiata* (see Figure 4.3).

It must be emphasized that DNA technology does not have the last word on what a species is. 'Species' is still ultimately a human construct. Many biologists express concern that DNA technology has led to a paradigm shift from the Biological Species Concept to the Phylogenetic Species Concept, and that the splitting of some species is not fully warranted.

Alpha, beta, and gamma species diversity

Species diversity is evaluated at three different scales referred to as alpha, beta, and gamma diversity (Figure 4.4). These are usually assessed simply in terms of species number (see species richness, Chapter 5).

Alpha diversity is the diversity of species occupying individual localities, habitats, or ecosystems.

Beta diversity reflects the rate, magnitude, and direction of change in the biodiversity between two or more different habitats or communities. It is usually expressed as a comparison of the total number of species unique to each of the areas.

Gamma diversity, also called regional diversity, refers to the biodiversity within a region and is usually expressed as the total number of species in the different ecosystems within that region.

These three types of species diversity are used to measure and monitor biodiversity at three different spatial scales. Alpha diversity refers to species diversity in one locality within a geographic region; it is usually

Figure 4.3 *Aurelia aurita*—just one of the seven species of Moon Jellyfish now recognized as a result of DNA sequencing.

Figure 4.4 Alpha, beta, and gamma species diversity: α-diversity in three separate localities, β-diversity between any two of the localities, and γ-diversity for the whole region which contains the three localities.

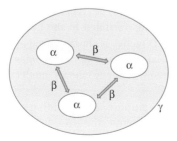

Image: © Michael Kent

expressed simply by the number of species (i.e., species richness). Beta diversity is a measure of the difference between species diversity in two or more localities; it is often used to compare two different habitats within a geographic area. Gamma diversity refers to the species diversity for a whole geographic area.

Ecosystem diversity

According to the Ecological Society of America, ecosystem diversity of a region refers to '… *all the different habitats, biological communities and ecological processes, as well as variation within individual ecosystems*'.

In the context of the marine environment, ecosystem diversity is the variation found in the ecosystems of a part of the global ocean or the variation in the ecosystems found in the whole global ocean.

In many assessments of ecosystem diversity, habitat diversity is a key determinant. Generally, the more habitats there are in an ecosystem, the higher the ecological diversity. Large stretches of the abyssal plain appear to be homogeneous with few habitats, resulting in low ecosystem diversity. By contrast, a rocky shore with a large tidal range is highly heterogeneous and will usually have a large number of habitats and high ecosystem diversity.

A closer look at marine habitats

'Habitat' has many meanings. In a non-scientific context, it is often used interchangeably with 'environment' to refer loosely to the surroundings of a person, a plant, or an animal.

In ecology, habitat is generally defined either in relation to the locality in which an organism lives or in relation to a particular assemblage of organisms together with their abiotic (non-living) environment.

In pre-University courses, a student needs to be able to distinguish the term habitat from other ecological terms with which it is commonly confused, such as environment, ecosystem, and niche. In this context, defining the habitat in terms of the place in which an organism lives is all that is required.

In habitat classification systems used as tools for ecological management and legislation, more precision in the definition is needed. Here a habitat is usually defined in terms of its physical features and by the assemblage of species living there. For example, the Joint Nature Conservation Committee (JNCC) for Britain and Ireland has devised a classification system that forms the basis of marine habitats included in the European Habitats Directive. One habitat listed in the Directive is characterized by the presence of *Mytilus edulis* (common mussels), *Semibalanus balanoides* (acorn barnacles), and *Patella vulgata* (limpets), on rocky shores moderately exposed to wave action (Figure 4.5).

Although fit for its purpose of providing descriptors for habitats included in the Habitats Directive, the JNCC habitat classification system is

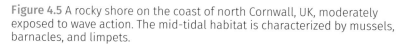

Figure 4.5 A rocky shore on the coast of north Cornwall, UK, moderately exposed to wave action. The mid-tidal habitat is characterized by mussels, barnacles, and limpets.

Photo: © Michael Kent

only one of many classifications used throughout the world. Each system differs in detail and is a compromise between different opinions. In other words, habitat classifications, and the definitions of habitat on which they are based, are subjective and imprecise. At the time of writing, attempts to create a unified and global marine habitat classification system have been unsuccessful.

Functional diversity

Functional diversity is a component of biodiversity concerned with the range of things organisms do that affect ecosystem processes.

There are various ways of measuring functional diversity but most depend on identifying functional traits: ecologically significant characteristics that make one group of organisms functionally distinct from another group.

Measurements of functional biodiversity are commonly used as indicators of the ecological health and resilience of an ecosystem: an ecosystem with a high functional diversity is generally assumed to be healthier and more resilient than one with a low functional diversity.

Whereas the use of species diversity as an indicator of health and resilience assumes all species have an equal effect on ecosystem processes, functional diversity does not. The disproportionate effects of some species are discussed in Case study 4.2 in which the removal of a keystone species has a dramatically greater effect on the functioning of ecosystems than the removal of other species.

Case study 4.2
Some species are more equal than others

Different species carry out different roles within an ecosystem—and some are more crucial to the overall balance than others. The way these interactions work can be clearly seen in the marine life balance at Mukkaw Bay on the Pacific Coast of the US.

When Robert T. Paine crowbarred a starfish off the rocks and hurled it 20 metres into the sea, he set in motion one of the most influential field experiments in the history of ecology.

It was 1963. Paine was an Associate Professor leading a group of students on a field trip at Mukkaw Bay on the Pacific Coast. When he came across a colourful community of intertidal creatures on boulders in the Bay (Figure A), he immediately realized that was just what he had been looking for to test the green world hypothesis, a radical new hypothesis that he had heard about when he was a student at the University of Michigan.

The hypothesis proposes that the world (or at least its terrestrial part) is mostly green because carnivores prevent herbivores from destroying all plants. The general consensus at the time was that the number of primary producers at the base of a food pyramid limits the number of herbivores. In turn, the number of herbivores limits the number of predators that feed on them, and so on up the food pyramid.

In other words, an ecosystem is subjected to 'bottom up control' with each level in a food pyramid being regulated by the amount of food in the level below it. But according to ecologists Nelson Hairston, Frederick Smith, and Lawrence Slobodkin, who proposed the green world hypothesis, this view doesn't explain why herbivore populations don't continue to grow until they eat all the plants. They suggested that the herbivores must be controlled not only from the bottom up, but also from the top down by predators.

Their suggestion turned the understanding of how ecosystem structure is controlled literally upside down. If true, it would have huge implications for ecological management and conservation measures. However, the suggestion was only theoretical and had little reliable scientific evidence to support it.

The potential importance of the hypothesis and the fact that there was insufficient evidence to support it excited Paine. When he saw the intertidal community at Mukkaw Bay, he realized in an instant that it was just what he needed to test the hypothesis.

He started the field experiment by identifying the prominent species that occupied the primary space (bare rock, or rock covered only with encrusting red algae) and working out their feeding relationships. He identified fifteen species prominent to the naked eye, with the pink starfish as the top predator. Figure B shows the types of organisms taken and eaten by the starfish *Pisaster ochraceus*.

Figure A A rocky shore community near Mukkaw Bay.

Photo: © Dave Cowles at inverts.wallawalla.edu.

Figure B Part of the rocky shore community studied by Robert T. Paine at Mukkaw Bay, showing the types of prey organisms taken by the top predator, the starfish *Pisaster ochraceus*.

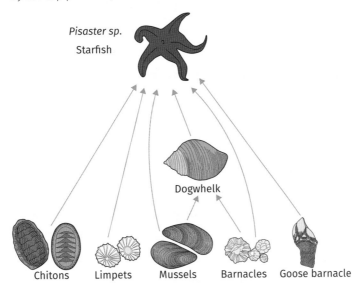

Source: Robert T. Paine

After establishing the feeding relations, Paine removed starfish from one rocky outcrop and threw them into deep water, while leaving starfish on other outcrops to continue their predation. Because there are always starfish marching onto the shore from deep water, he had to remove them repeatedly during the summer months so that he could compare what happens to the rocky shore ecosystem with and without the top predator.

Within 18 months, Paine's intervention had such a dramatic effect on the ecosystem that he had (in his own words) 'hit ecological gold'. In the area in which the top predator had been removed, the number of species decreased from 15 to 8. After three years, the number went down to 7. By another seven years, the site cleared of starfish was almost a monoculture of mussels. The mussels had taken over almost all the available primary space, the limiting resource of this ecosystem. Contrary to the opinion at the time, the top predator increased species diversity by preventing mussels occupying most of the primary space.

Paine described *Pisaster ochraceus* as a **'keystone species'** because it played a role similar to that of a keystone in an arch. If you pull out the keystone of an arch, the arch collapses; if you remove a keystone species from an ecosystem, the ecosystem collapses. Paine had shown that a top predator could have a huge impact on an ecosystem, far beyond just affecting the species it preys upon.

Based on the example of *Pisaster ochraceus* at Mukkaw Bay, Paine defined a keystone species as one whose population is the 'keystone of the community's structure', whereby the integrity and stability of the community are determined by its activities and abundance. He argued that a keystone species has a disproportionate influence on key community properties and claimed that variation in the abundance of other predators 'would produce no impact comparable to that produced by variations in the keystone species'.

? Pause for thought

Suggest why Paine's keystone species concept has been criticized for being 'broadly applied' and 'poorly defined', and that basing conservation strategies on keystone species might be dangerous.

Throughout the 1970s and 1980s, a range of species was identified as keystone species. One of these was the sea otter *Enhydra lutris* (Figure 4.6), a keystone species of kelp ecosystems along the Aleutian Islands in the north Pacific Ocean.

By feeding on sea urchins, sea otters maintain the intertidal and sub-tidal community structure. The importance of sea otters was demonstrated in areas in which they had become wiped out by orcas (killer whales) (Figure 4.7).

Orcas' natural prey is mainly other whales. But after World War II, the whaling industry had grown to such an extent that whales became almost extinct in parts of the Pacific. Killer whales resorted to other prey: first seals and sea lions were targeted, then sea otters.

In areas deprived of sea otters, species diversity was low and primary production of kelp was significantly reduced. Restoring sea otters to kelp beds increased species diversity and primary production; the kelp beds thrived.

Figure 4.6 A sea otter eating a sea urchin.

Photo: Mike Baird from Morro Bay, USA

Figure 4.7 Orca, a fierce predator of whales and other marine mammals.

Photo: Pixabay

Trophic cascades

Robert T. Paine, working with James Estes (the scientist who discovered the connection between sea otters and orcas), referred to the top-down effects of sea otters on the kelp ecosystem as a trophic cascade. In general terms, a trophic cascade occurs when a top predator controls the distribution of resources in an ecosystem and its removal leads to many effects, some direct and some indirect. When the voracious and deadly orcas began feeding on sea otters, they added a new trophic level to the kelp ecosystem, causing a trophic cascade in the three levels below them (Figure 4.8).

Figure 4.8 Changes in sea otter abundance (a) over time at several islands in the north Pacific and changes in (b) sea urchin biomass, (c) grazing intensity, and (d) kelp density measured from kelp forests. The proposed mechanisms of change are portrayed in the drawings: those on the left show how the kelp forest ecosystem was organized before the sea otter's decline; those on the right show how this ecosystem changed with the addition of killer whales as a top predator. Heavy arrows represent strong trophic (feeding) interactions; light arrows represent weak interactions.

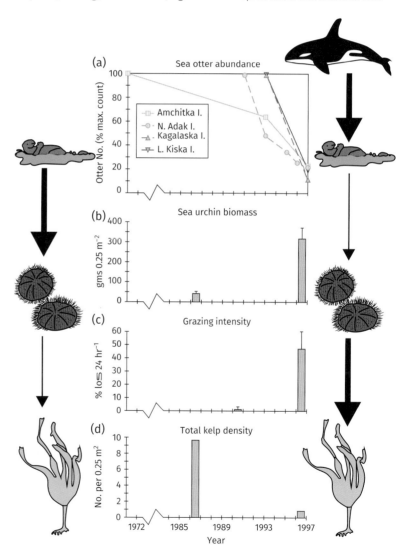

Scientific approach 4.1
Biodiversity monitoring and value for money

So far, we have seen that biodiversity is a multidimensional concept that is difficult to explain, and (as we will see in the next chapter) it is just as difficult to measure. Merely acknowledging the theoretical difficulties associated with aspects of biodiversity is not very useful for ecologists and biological conservationists tasked with monitoring biodiversity changes within ecosystems. They are confronted with the practical problem of choosing a component to focus on. Ecological monitoring can be expensive in terms of both time and money. Funding bodies want to make sure they get value for their money.

Figure A is a key figure in a handbook by the Group on Earth Observation (GEO), a voluntary international partnership of 102 governments and 92 participating organizations which share a vision of a future in which decisions and actions for the benefit of humankind are informed by coordinated, comprehensive, and sustained Earth observations.

The figure relates to different ways of monitoring changes in biodiversity. It shows that biodiversity embraces differences in composition, structure, and function at several levels of biological organization, from the biomolecular to the biosphere. The authors of the GEO handbook note that there is no single 'right' way to monitor and measure biodiversity. They state that, ideally,

Figure A Biodiversity—the variety of life in all forms and at all levels.

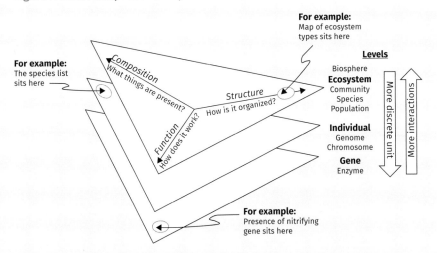

Source: © Walters, M. and Scholes, R.J. 2017

those tasked with monitoring biodiversity should capture elements of all components at all levels, and '... *be able to move seamlessly between them*'. However, the GEO recognizes that this ideal is not very practical, and that in any particular situation there will inevitably be stronger emphases on some components and some levels rather than on others.

Traditionally, biodiversity monitoring has been focused on community composition at the species level. The GEO advises that those planning ecological monitoring should be guided primarily by the information needed by the end users (e.g., the managers of a marine park); secondly by what can be monitored using available technology; and only then by what had been collected in the past.

The GEO document notes that as monitoring shifts from the ecosystem towards the organism and ultimately the gene, the entities dealt with become more focused and precise, but that this comes at a price: information is lost about the interactions between entities and the properties that emerge from those interactions.

❓ Pause for thought

The GEO handbook states that '... *having a deep understanding of biodiversity was an essential element for survival for most of the human past*'.

Is this statement justified by any prehistorical evidence?

Chapter summary

- Following the 1992 Rio Earth summit, the United Nations General Assembly instituted International Biodiversity Day as an annual celebration, and to increase understanding and raise awareness of biodiversity issues. In 2012, the Day was dedicated to Marine Biodiversity. The Summit also led to the Convention on Biological Diversity, probably the most influential biological conservation document.
- The term 'biodiversity' is a value-laden term because of its links to biological conservation.
- Marine biodiversity can be studied at any level of biological organization, from biomolecules to the biosphere. Whether or not viruses should be included as part of biodiversity depends on how life is defined.
- Genetic diversity includes the variation of genetic information in individuals, species, populations, and the biosphere. High genetic diversity is associated with high ecosystem resilience.
- Species diversity is the most common level of biological organization at which marine biodiversity is assessed. The difficulty of defining and naming species is considered.

- Marine ecosystem diversity is considered in relation to marine habitats and the importance of keystone species.

 ## Further reading

Ehlers, A., Worm, B., and Reusch, T. (2008) Importance of genetic diversity in eelgrass *Zostera marina* for its resilience to global warming. *Marine Ecology Progress Series*, 355, 1–7. 10.3354/meps07369.

An interesting insight into how genetic diversity can be used to measure the effect of climate change on an ecosystem.

Estes, J.A., Tinker, M.T., Williams, T.M., and Doak, D.F. (1998) Killer Whale Predation on Sea Otters Linking Oceanic and Nearshore Ecosystems. *Science*, 282, 473–475.

Available at www.sciencemag.org. An excellent, accessible primary source of information about the link between sea otters, killer whales, and kelp ecosystems.

Spicer, J. (2021) *Biodiversity: a beginner's guide*. Oneworld.

An engaging, authoritative introduction to biodiversity written by a professor of marine biology for students and the general reader.

Walters, M., and Scholes, R.J. (2017) *The GEO Handbook of Biodiversity on Observation Networks*. Springer.

A 'how-to' book on observing and monitoring biodiversity aimed mainly at technical specialists, but useful to anyone wanting to know more about practical aspects of biodiversity.

Wilson, E.O., and Peter, F. (eds) (1988) *Biodiversity*. The National Academy of Sciences.

One of *the* classic books on biodiversity. Rather dated, but still worth reading and a good resource for early references on biodiversity.

 ## Discussion questions

4.1 Discuss why there is no comprehensive, rigorous, universally accepted concept of species.

4.2 Suggest why ecologists tend to be 'lumpers', assigning organisms with slight variations into one species, whereas evolutionary biologists tend to be 'splitters', assigning the same organisms to different species.

4.3 About 200 years ago, the Native American Chief Seattle talking about how we are connected to life on Earth stated that 'Humankind has not woven the web of life. We are but one thread in it. Whatever we do to the web, we do to ourselves. All things connect'. Assuming we are 'one thread' in the kelp ecosystems of the north Pacific, discuss whether or not we could be regarded as a top predator, key species, an invasive species, or any combination of these.

5 MEASURING MARINE BIODIVERSITY

'All science is either physics or stamp collecting.'

This quote is attributed to New Zealander Ernest Rutherford (Figure 5.1), the Nobel Prize winning scientist who established the nuclear structure of the atom and was, famously, the first person to split the atom.

Rutherford is one of the greats of science. His rather disparaging statement is most likely directed, not at chemists (his Nobel Prize was, after all, in chemistry) or biologists, but at all those who conduct science by observation and speculation, rather than by measurement.

In this section we are going to look at how biodiversity research is carried out scientifically by being quantified. Cataloguing, whether types of stamps or types of organisms, has its importance (see Case study 5.1). But by measuring marine biodiversity, or at least some aspects of it, ecologists can postulate hypotheses and make predictions which can be subjected to rigorous scientific testing. There are many different ways of measuring biodiversity: species diversity, genetic diversity, ecosystem diversity, and functional diversity are just some of them, and they all have value. In this chapter we will look at each of these measures of biodiversity in turn, with our focus on the marine environment.

Figure 5.1 One of two New Zealand stamps issued in 1971 to mark the centennial of the birth of Ernest Rutherford.

Image: svic/shutterstock

Measuring species diversity

It is at the level of species that marine biodiversity is most commonly measured. All measurements of the diversity of species involve identifying and counting the number of species in the locality under investigation. But before any such measurements are made, it is necessary to understand what a species is.

A species is commonly defined as a group of organisms with similar features which can interbreed to produce fertile offspring. This definition works well for organisms which reproduce sexually, but the definition cannot be applied to asexually reproducing organisms, or where reproductive behaviour has not been observed, for example in extinct species.

In practice organisms are usually given species status on the basis of morphological features, but biologists often have difficulty deciding which features to use, or how similar two individuals have to be to belong to the same species. In addition, many species, such as the dogwhelk (*Nucella lapillus*), are polymorphic (see Figure 5.2).

Intraspecific variation is an important contributor to biodiversity. Different morphs usually have different tolerances to environmental factors such as predation pressure or thermal factors. For example, on the north Atlantic coast of the USA, darker dogwhelks are more affected by heat stress than lighter-coloured dogwhelks.

Science writer Kevin Zelnio summed up the species situation really well in a *Scientific American* article, when he wrote 'The species concept "problem" has pervaded for many years and will not be resolved anytime

Figure 5.2 A range of colour morphs of the dogwhelk (*Nucella lapillus*) from a rocky shore on the coast of north Cornwall.

Photo: © Michael Kent

soon, if ever. The problem, of course, being that no two scientists will agree on universal definitions of what the darn things are!'

Wherever possible, several criteria should be used to define a species (see also The bigger picture 5.1).

Species richness

Species richness is the simplest measure of biodiversity. It is most commonly expressed as the number of species observed in a community, locality, or during a particular survey. It can also be calculated as an index which recognizes that measures of species richness depend on sample size. For example, Menhinick's diversity index (DMn) measures species richness as number of species observed (S) divided by the square root of the number of individuals (N) in the sample:

$$DMn = S / \sqrt{N}$$

Menhinick's diversity index remains strongly influenced by sampling effort.

Unlike species diversity indices, such as the Simpson Diversity Index which we will consider later, measurements of species richness take no account of the number of individuals belonging to each species.

On its own, species richness has serious limitations. It assumes that all species are of equal importance in an ecosystem although this is patently not the case in most situations (see Case study 4.2). Nevertheless, species richness may be the only practical measurement of biodiversity and it is an extremely valuable starting point for projects, such as the Census Of Marine Life (COML), designed to discover new species.

The Census of Marine Life

The COML was an 11-year international collaboration that ran from 2000 to 2010. It was one of the most ambitious marine science projects ever undertaken. Its primary aim was to assess the global diversity of marine life, its distribution, and its abundance, and it has made an enormous contribution to our understanding of marine biodiversity.

More than 2700 scientists, from over 80 nations, were involved in COML. About 540 expeditions were undertaken during the project; it resulted in the discovery of more than 6000 potential new species, including the unusual Yeti Crab (Figure 5.3).

The Yeti Crab is an incredible animal. It was found at a depth of more than 2000 metres on a hydrothermal vent at a site near Easter Island.

Figure 5.3 The Yeti Crab (*Kiwa hirsuta*), one of the many thousands of new species discovered during the Census of Marine Life project.

Photo: Oregon State University/Flickr

Although the crab was observed feeding on mussels, it is probably omnivorous. Its profusion of hair-like setae harbour chemosynthetic bacteria; these may provide an additional source of nutrients for the crab.

The Yeti Crab is related to lobsters, shore crabs, and shrimps. But it has a unique combination of features that include the hair-like setae, lack of eyes, and an unusual genome. These features make it difficult to place into any known family. Consequently, *Kiwa hirsuta* has been described as representing not only a new species, but also a new family of crustaceans.

The COML project produced the most comprehensive inventory of marine life ever compiled and catalogued, generating more than 2600 scientific publications and more than 30 million distribution records by 2010—and more have been added since. In addition to describing new species, the publications documented long-term and widespread declines in marine life, as well as encouraging examples of resilience in some marine ecosystems.

This truly amazing project has provided a baseline picture of global marine life that is being used to forecast, measure, and understand changes in marine environments, as well as being used to inform management and conservation decisions.

How was it done? The COML scientists used a combination of traditional taxonomic techniques and modern procedures to identify and count marine life. However, environmental DNA (eDNA), a new and innovative technique that is becoming an increasingly valuable tool for measuring species richness, was not available to COML scientists (see Scientific approach 5.1).

Scientific approach 5.1
eDNA changes everything

In a 2018 *National Geographic* article, Jesse Ausubel, one of the leaders of the COML project, said that if eDNA had been available when the project was conducted it would have been done faster, better, and more cheaply. Asubel added, 'It changes everything in marine science'.

What is eDNA?

eDNA is a technique derived from DNA fingerprinting (now known as DNA profiling, or **DNA barcoding**) invented in 1984 by Professor Sir Alec Jeffreys as a forensic tool. Just as each person at the scene of a crime may leave behind DNA with its own base sequence that can be barcoded and used to identify the individual, each organism in an environment leaves behind DNA that can be barcoded and used as evidence of its presence. A key component of the method is extracting a short section of DNA from a specific gene or genes.

eDNA barcoding involves taking a sample, such as some seawater, from the environment, extracting the DNA, treating the DNA with an appropriate **DNA primer**, amplifying the products by **Polymerase Chain Reactions** (**PCR**) and then analysing the sequence of nucleotide bases of DNA and displaying them as bands in a barcode (see Figure A).

A key part of the process is comparing the barcodes in the sample DNA with DNA barcodes of known species in a reference library. The number of species that can be identified is limited to those contained within the reference library.

Environmental DNA barcoding can reveal which species occur in a marine habitat. All it requires is a sample from the habitat as small and simple as a cup of water. eDNA barcoding relies on bits of shed skin, mucus, faeces, or body fluids that organisms leave behind in the environment. Analyses of the sequence of DNA in these bits can be used to identify the species that have been present before or at the time of sampling.

Among the most useful sources of environmental DNA are sponges (see Figure B), which are being used as natural eDNA samplers. Sponges sift up to 10 000 litres of seawater a day, trapping and concentrating bits of DNA from the organisms that swim around them. Researchers claim that sponges filter seawater much more effectively than any artificial device.

The ability to identify organisms by bits of DNA left behind in the environment is a major advance in marine ecological research. eDNA techniques have become so accessible that they are being used in Citizen Science projects. For example, in the spring and summer of 2017, a group of New York high school students collected samples of water on a weekly basis from the fishing pier at Coney Island. eDNA barcoding provided evidence that

Figure A DNA barcodes in three marine crustacean species: *Scina crassicornis* (an amphipod), *Mollicia tyloda* (an ostracod, sometimes called a sea shrimp), and *Sapphirina metallina* (a copepod). The barcodes are derived from the base sequence in a single gene, the mitochondrial cytochrome oxidase I gene (COI), that is known to occur in all marine zooplankton but which varies greatly from one species to another.

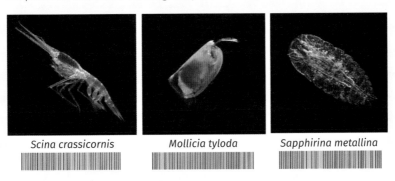

Scina crassicornis   *Mollicia tyloda*   *Sapphirina metallina*

Source: Ann Bucklin, CC BY-ND

Figure B An Azure Vase sponge, *Callyspongia plicifera*, a species found in the Bahamas.

Photo: Focused Adventures/Shutterstock

34 species of marine life, including sharks and rays, had been in the waters around the pier during that period.

❓ Pause for thought

- Using eDNA barcoding as a tool in marine ecological research could have major environmental and economic implications. Suggest what these might be.

- In a review on DNA barcoding (see Further reading), Ann Bucklin and co-authors said that 'DNA barcoding will complement—not supplant or invalidate—existing taxonomic practices'. What do you think?

Although eDNA technology enables almost anyone to be able to identify an organism, it does not make the role of taxonomy any less important. If anything, it makes it more important.

Like many other biological terms, the precise definition of taxonomy varies from one authority to another, but essentially it entails naming, defining, and classifying organisms on the basis of shared characteristics. Traditionally, those characteristics were morphological. Nowadays, genetic characteristics are increasingly used. Nevertheless, most specialist taxonomists (scientists trained in taxonomy) use a combination of methods from the molecular and morphological to the ecological.

When an organism new to science is discovered, whoever found it can name it using the rules of nomenclature, and a molecular biologist can map its genome with relative ease. However, it takes someone trained in

Figure 5.4 A Venn diagram showing that nomenclature and taxonomy have separate domains except in relation to the type specimen, as this is designated through the rules of nomenclature that embed scientific names into whatever is the current classification of the biological world.

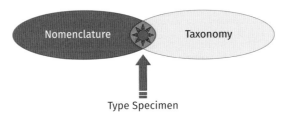

Type Specimen

taxonomy to give it a taxonomic status, assign it to a species, and place it in a classification system.

The importance of taxonomy is the theme of a 2018 paper written by Scott A. Thomson, with dozens of other co-authors from many nations (see Further reading). The paper stated in no uncertain terms that '*The critical importance of taxonomy and the taxonomic process in the global quest to mitigate biodiversity loss cannot be overemphasized*'.

The paper included a diagram that showed that the domains of nomenclature and taxonomy overlap only in relation to the type specimen (Figure 5.4), the specimen used for naming and describing a species or subspecies.

There is an ongoing argument between conservation biologists who want taxonomy to be more strictly governed, and taxonomists who do not, and we could spend pages discussing the fine distinctions between systematics and taxonomy. But there is a real concern, shared by many professional ecologists and biological conservationists, that we do not have enough scientists trained in taxonomy to deal with biodiversity issues. In short, we need to know what we have, so we can recognize what we are losing.

Way back in 1995, the authors of a US National Research Council report entitled *Understanding Marine Biodiversity* included this explanation of the critical role of taxonomy: '*Changes in the sea caused by anthropogenic effects are most commonly measured by changes in the distribution and abundance of species. The loss of individual populations of species may affect the genetic diversity of a species, and thus impact the survival of the species itself. The loss of ecosystem diversity restricts the habitat available for a species, and so, too, may affect the species' survival. At the center of these cascading effects is the species. The ability to identify individual species is thus the key that permits the opening of the first door to an understanding of community structure and function. And yet for all marine systems, the ability to "simply" identify the species present is now threatened by a continuing loss of scientists with the knowledge and ability to understand and describe biodiversity. Moreover, in many systems, species diversity is so poorly known—that is, so many species and entire groups of higher taxa remain undescribed—that the impact of human activities on diversity is difficult to assess at all.*'

The situation today is no better. In fact, it is probably worse, as we are finding so many more new species that need to be identified, named, and classified. The need for more biologists trained in taxonomy is echoed in the Thomson et al. paper; it suggests that the argument between conservationists and taxonomists reflects a misunderstanding of the scientific basis of taxonomy and nomenclature, and the relationship between them. The concluding paragraph in the paper includes a statement that echoes that in the 1995 paper: 'The critical importance of taxonomy and the taxonomic process in the global quest to mitigate biodiversity loss cannot be overemphasized'.

Species diversity indices

If you are a reader taking a pre-University course in a biological or environmental science, you will probably be familiar with the Simpson diversity index; it is commonly used to measure biodiversity in marine habitats. There are several ways of expressing the Simpson diversity index (D) such as

$$D = N(N-1) / \sum n(n-1)$$

where
n = total number of organisms of a particular species, and
N = total number of individuals of all species

D can range from 0 (no diversity) to 1 (infinite diversity).

The use of the Simpson diversity index assumes the areas sampled (e.g., 1m^2 quadrats) are located randomly or systematically. No matter how it is expressed, the index takes into consideration both the number of species and the total number of individuals. A community dominated by one or two species will have a low index, and be considered less diverse, than one with several different species and the same abundance.

One weakness of the Simpson diversity index is that it makes no allowance for differences in size of individuals: one Blue Whale is equivalent to one marine bacterium! Also, it may be difficult to calculate the Simpson diversity index of colonial organisms, such as coral, because it is not always easy to know what constitutes an individual organism.

There are many different species diversity indices; some are based on biomass rather than individual numbers but, unlike species richness, they all take into consideration the abundance of individuals belonging to each species.

Species evenness

Species evenness focuses on the relative abundance of different species in a sampling area. It is a measure of how close in numbers each species in a community is.

There are various ways of calculating species evenness, but detailing them goes beyond the scope of this book. Suffice it to say that a measurement of species evenness would show that a rock pool community containing 3 fish, 4 sea anemones, and 4 brittle stars would have a higher species evenness than a rock pool with 1 fish, 9 anemones, and 1 brittle star even though the two rock pools have the same number of species and the same total number of individuals and would therefore have the same Simpson diversity index.

Measuring genetic diversity

There are numerous ways of measuring genetic diversity. One well-tried and tested method is by calculating the heterozygosity index (H) (Figure 5.5), a measurement of the proportion of heterozygotes in a population:

H = number of heterozygotes / number of individuals in the population

Until recent advances in DNA technology, one of the commonest methods for measuring heterozygosity in marine organisms involved using gel electrophoresis coupled with histochemical staining of specific proteins. The technique is relatively inexpensive and provides a measure of many variable and non-variable genes in individuals. Gel phenotypes are relatively easy to interpret and there are computer programs available for data analyses. Genetic diversity measured using gel electrophoresis can be expressed as the average heterozygosity over many gene loci coding for proteins, or the proportion of the individuals that are heterozygous at a single gene locus.

It is assumed that genetic diversity measured by protein heterozygosity is positively correlated with ecological fitness—the more genetic

Figure 5.5 The heterozygosity index, Hi, calculated for four gene loci of one shore crab, and Hp, calculated for one gene locus of a population of shore crabs.

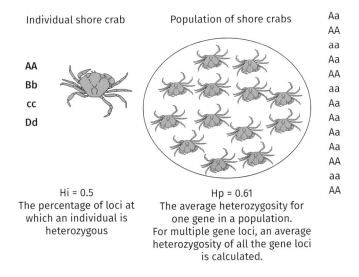

Individual shore crab   Population of shore crabs   Aa
                                                                  AA
                                                                  aa
                                                                  Aa
AA   AA
Bb   aa
cc   Aa
Dd   Aa
                                                                  Aa
                                                                  Aa
                                                                  AA
                                                                  aa
                                                                  AA

Hi = 0.5   Hp = 0.61
The percentage of loci at   The average heterozygosity for
which an individual is   one gene in a population.
heterozygous   For multiple gene loci, an average
                               heterozygosity of all the gene loci
                               is calculated.

diversity, the more likely the individual or population is to survive. There is some evidence to support this assumption. For example, protein heterozygosity is positively correlated with ecologically important functional traits such as growth rate and disease resistance. But the evidence is not conclusive.

Another weakness of using protein heterozygosity as a measure of genetic diversity is that the genes coding for proteins account for only about 10% of a typical genome. Therefore protein heterozygosity may not provide an accurate representation of the genetic diversity of whole genomes.

Whichever method is used to measure genetic diversity, it is usually coupled with computers programmed to deal quickly with the genetic information generated, so that genetic diversities can be analysed and visualized rapidly and effectively.

Measuring ecosystem diversity

Ecosystem diversity is the number and distinct kinds of ecosystem types in a defined area, their spatial patterns, and degree of similarity.

Ecosystem diversity is often used as evidence in proposals for designating protective status to a marine area. For example, in 2019 Plymouth Sound was designated the UK's first National Marine Park (see Figure 5.6). The Sound boasts incredible rocky and sandy shores, and probably the greatest concentration of different types of substrate of any estuary in the UK. The success of the proposal for the Park was at least partly due to its high marine ecosystem diversity and descriptions of its many and varied coastal and estuarine ecosystems.

Qualitative evidence of marine ecosystem diversity, such as this, is valuable, but quantitative evidence is much more convincing. Unfortunately, although ecosystem diversity is often referred to in many scientific and general publications, it is rarely defined precisely or measured.

Figure 5.6 A view of Plymouth Sound, the first UK National Marine Park.

Photo: Seb Perrotin/Flickr

Measurements of marine ecosystem diversity depend on a general agreement as to what, precisely, an ecosystem is. However, as we have discussed in Chapter 1, 'ecosystem' is often used synonymously with 'habitat', and the definition of both vary according to context and purpose.

Measuring functional diversity

As mentioned in Chapter 4, functional diversity is the component of biodiversity generally concerned with the range of things that organisms do which affect ecosystem processes.

Functional diversity is sometimes used simply to refer to the presence or absence of specific functional groups. A functional group or guild is a group of species that perform the same role in an ecosystem. For example, as shark and tuna both eat fish, they belong to the piscivores (the fish-eating group). In contrast, some of their prey, being found lower in the food web, belong to the planktivores (the plankton-eating group). An ecosystem that includes both piscivores and planktivores would have greater functional diversity than an ecosystem that has only one of these groups.

Measuring functional diversity by counting the number of functional groups represented by the species in a community has significant problems. For example, many species can belong to more than one functional feeding group at different times of their lives, in different seasons, or even at different times of the day. Most researchers use measures that do not require species to be grouped. Instead, they base their measurements of functional diversity on functional traits.

A functional trait is a measurable structural, physiological, or behavioural characteristic of an individual organism that has significant effects on the organism's responses to the environment and on the functioning of an ecosystem. Functional traits usually include some aspect of how an organism moves, feeds, reproduces, or the time of year it is active.

Reliable estimates of functional diversity depend on choosing meaningful, ecologically significant traits for the scale and level of diversity being measured. Measurable functional traits chosen to calculate functional diversity in marine phytoplankton include those related to cell size and shape, tolerance and sensitivity to environmental conditions, motility, the ability to fix nitrogen, and requirements for silicon and iron. Although these functional traits might enable reliable functional diversity to be calculated for pelagic phytoplankton in specific localities, they might not be suitable for benthic phytoplankton.

There is no universally accepted list of functional trait descriptors that covers all marine organisms in all marine environments at all levels and scales. However, BIOTIC (the Biotic Traits Information Catalogue), initiated by the Marine Biological Association, provides scientists working within the field of benthic community ecology with a tool for choosing functional traits. The BIOTIC database contains information on over 40 biological trait categories on selected benthic species, primarily benthic invertebrates and algae.

Measuring functional diversity in real ecosystems can be a complex process.

According to a highly regarded review by Owen Petchey and Kevin Gaston (see Further reading), measuring functional diversity requires, ideally, each of the following:

1. Appropriate functional information (traits) about organisms to be included in the measure, and irrelevant information to be excluded. It is generally agreed that this is the most critical part of the process.

2. Traits to be weighted according to their relative functional importance.

3. The statistical measure of trait diversity to have desirable mathematical characteristics.

4. The measure to be able to explain and predict variation in ecosystem processes.

To achieve all these requirements, a measurement of functional diversity involves some very complex statistical and mathematical procedures. Here we will confine ourselves to a simple hypothetical example to show how functional diversity can be measured on a rocky shore with two different communities, without going into the statistical and mathematical details.

Imagine that in community 1 we find four species in a random quadrat: *Mytilus edulis* (common mussel), *Patella vulgata* (common limpet), *Chthamalus montagui* (a southern acorn barnacle), and *Semibalanus balanoides* (the northern acorn barnacle). And in community 2 we find four different species in a quadrat: *Henricia sanguinolenta* (a blood-red starfish), *Patella ulyssiponensis* (China limpet), *Pomatoceros triqueter* (a polychaete tubeworm), and *Rhodymenia holmesii* (Holme's rose, a red seaweed).

Figure 5.7 Simple representations of species richness in community 1 and community 2. The squares are delimited by lines of equal length, indicating that the species are regarded as being of equal value.

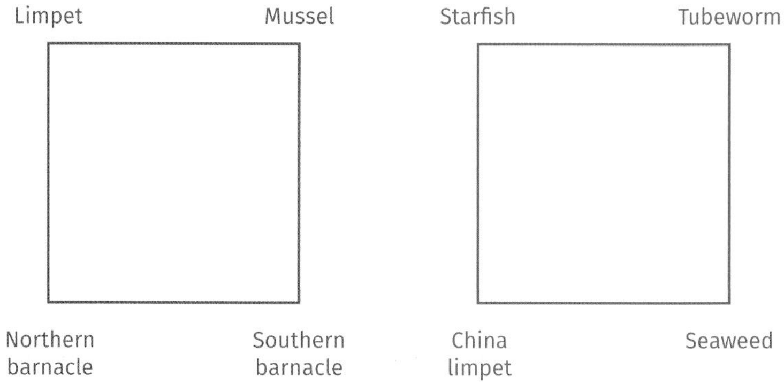

If we were to use species richness as our measure of biodiversity, we would have to conclude that the two communities have an equal diversity of 4 species each (Figure 5.7).

The reason for the species richness of the two communities being the same is that each species in the two communities is considered to be equal and have the same impact on ecological processes. Using species richness also implies that the southern barnacle, for example, is as different to the northern barnacle as it is to the rose seaweed. However, even to the untrained eye community 2 actually looks more diverse than community 1.

When we use functional diversity instead of simple species richness as our metric of biodiversity, measurable functional traits allow us to score species along a continuum of dissimilarity. Some species share traits, such as being the same size and eating the same type of food, and are therefore more similar than those that do not. By applying functional traits to the two rocky shore communities, we can obtain polygons where the length of line between two species is related to their dissimilarity. Here, Figure 5.8 reflects, in an approximate way based on an understanding of the biology involved, the similarities between the organisms in each of the communities; the shorter the lines, the greater the similarity. In community 1, all the organisms are molluscs, and the barnacles and mussels are all filter feeders, thus the lines between species are short. In community 2, all the organisms are functionally distinct and belong to different taxonomic groups; there's a mobile predator, a grazer, a filter-feeder, and a photoautotroph, so the lines between species are longer. The smaller polygon for community 1 indicates that it has a smaller functional diversity than community 2, but it does not indicate with any accuracy how much smaller it is.

In scientific papers various mathematical and statistical procedures are used to construct diagrams such as dendrograms to depict quantitative differences between the functional traits of species. In their review paper, Owen Petchey and Kevin Gaston state that *'Measuring functional diversity is a problem of how to measure the amount of variation represented by a set of points in multivariate space'*.

In conclusion, functional diversity is a powerful and important component of biodiversity, but it is also rather complex. Measuring functional diversity

Figure 5.8 Graphical representation of the functional diversity in two rocky shore communities. The polygons show community 1 has a lower functional diversity than community 2.

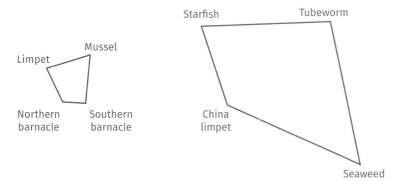

is probably of greatest value when it is used in conjunction with other measures of biodiversity, such as species richness.

No matter how it is done, measuring marine biodiversity involves obtaining samples from the marine environment. Whenever we go down onto the shore or into the sea to collect specimens for scientific analyses, or even for display in aquaria, we should be aware of the harm that we might do to the habitats we are visiting and the organisms we are collecting. The unintended consequences of our activities are considered in The bigger picture 5.1.

The bigger picture 5.1
Unintended consequences

Phillip Henry Gosse FRS (1810–1888) was a famous Victorian naturalist and author. His books on the sea and shore, illustrated with stunning colour plates (see Figure A), sparked a huge interest in marine life. For example, he described in detail the corkwing wrasse (*Crenilabrus melops*, sometimes called the sea-partridge) which is the smallest species of wrasse in British waters (see Figure A). This is the wrasse most commonly seen in rock pools; individuals may occupy the same pool throughout the year. Wrasses have the peculiar habit (for fish) of sleeping on their sides. Males and females are different sizes and colours. The male corkwing builds a relatively large, complex ball-shaped nest among the seaweed. He is highly territorial and guards the nest aggressively. This was a fascinating fish people could find and see for themselves.

After creating the first public aquarium at London Zoo in 1853, Gosse wrote an exceedingly popular and profitable manual on *The Aquarium*. It provided sufficient information for anyone to set up a home aquarium, such as the one illustrated in Figure B.

Figure A This plate from Phillip Gosse's book *A year at the shore* shows a brightly coloured male and a larger, greenish-brown female corkwing wrasse.

Figure B An illustration of a marine aquarium from *Ocean Gardens: The History of the Marine Aquarium* (1857) by Henry Noel Humphreys. No wonder people wanted to have their own!

A DESIGN FOR AN AQUARIUM MOUNTED IN HANDSOME RUSTIC-WORK.

Unfortunately, Gosse was so successful at enthusing his readers that they invaded British shores in their hordes to see marine wildlife for themselves, and to collect specimens for their marine aquaria. An unintended consequence of Gosse's ability to enthuse his readers is that they developed such an appetite for seashore visits and collecting that they destroyed much of what they had come to enjoy and understand.

In his book *Father and Son* (1917), Edmund Gosse, Philip's son, wrote about the specimen collecting expeditions on the Devon coast he made with his father in 1858, five years after the publication of *The Aquarium*. In the following passage, Edmund describes how the rock pools, once so full, had become empty:

> There is nothing, now, where in our days there was so much. Then the rocks between tide and tide were submarine gardens of beauty that seemed to be fabulous, ... if we delicately lifted the weed-curtains of a windless pool, though we might for a moment see its sides and floor paved with living blossoms, ivory-white, rosy-red, and amethyst, yet all that panoply would melt away, furled into the hollow rock, if we so much as dropped a pebble to disturb the magic dream.

He then goes on to explain poignantly how the rock pools

> ... thronged with beautiful sensitive forms of life ... became ... profaned and emptied, and vulgarized. An army of 'collectors' has passed over them, and ravaged every corner of them. The fairy paradise has been violated, the exquisite product of centuries of natural selection has been crushed under the raw paw of well-meaning, idle minded curiosity.

Unexpected consequences are not just a historical problem—there are many contemporary examples of how curiosity can kill the things we value.

In 2016, editors of the journal *Biological Conservation* rejected a scientific paper for publication because they regarded the research it described as '*unnecessary and inappropriate*'. The paper described an investigation to prove there are more fish inside marine protected areas than outside them— an investigation that had involved killing over a thousand fish. The editors decided this study only confirmed something already well established. But that was not the only reason for rejecting the paper. The researchers used gill nets, a destructive method of sampling which not only caught and killed the target fish, but also other species.

Before deciding whether or not to publish the paper, the editors asked the researchers whether they had considered using non-destructive sampling techniques. The researchers responded that non-destructive alternative methods were more expensive and time-consuming, and that they had all the necessary permits to carry out the sampling using gill nets. Not satisfied with the answers, the editors rejected the paper on ethical grounds.

In an article in *Biological Conservation* (see Further reading), the editors reported that they had '... *faced this type of decision on other papers, too, and have sometimes accepted them for publication because the impacts to wildlife were well justified; other times we rejected them because the value gained did not warrant the killing or harming of wildlife*'.

The article suggests how field investigators can minimize unintended consequences, arguing that conservation biologists, ecologists, marine biologists, and other scientists '... *should set the highest standards for carrying out fieldwork, especially when working within protected areas and with threatened species. Merely filling in the needed paperwork and getting official approval is not enough*'.

The authors propose the following 10 considerations for respectful conduct during biological field sampling:

Before sampling

1. Justify any potential adverse impacts of the research in terms of advancing scientific understanding.
2. Comply with the spirit of institutional and national regulations regarding research and responsible care and use of animals, collecting samples and specimens, and working in protected areas.
3. Apply the precautionary principle in assessing the potential impact of the research on species and their habitats. These potential impacts include inadvertent transport of pests, pathogens, and introduced species.

During sampling

4. Avoid killing animals and plants, especially species of conservation concern and species in protected areas.

5. Minimize disturbance to wildlife and habitats. Ensure that accidentally captured animals will be carefully and immediately released alive.

6. Minimize stress to animals that are sampled or handled.

After sampling

7. Remove research equipment and materials from study sites.

8. Maximize future benefits of research by archiving samples for future research and educational use.

9. Promptly report information that responsible authorities should know about, such as pollution and rare and invasive species observations.

10. Publish findings and data in publicly accessible permanent archives for use in future research, education, and management. And whenever possible, inform the local community about the results through popular articles and public talks.

❓ Pause for thought

The editors of *Biological Conservation* rejected a paper describing a study undertaken on fish inside and outside a Marine Nature Reserve for ethical reasons.

1. What 'non-destructive sampling techniques' might the researchers have used to compare fish inside and outside marine protected areas?

2. Suggest how and why the editors' decision might have differed if the animals under investigation were pelagic prawns rather than fish.

⊜ Chapter summary

- By measuring a component of marine biodiversity, ecologists can postulate hypotheses and make predictions that can be subjected to rigorous scientific testing.

- Species diversity is the component of marine biodiversity most commonly measured. It can be measured by species richness, a species diversity index, or by species evenness. Environmental DNA is transforming the way species diversity is measured in the marine environment.

- Genetic diversity is measured in a variety of ways, but one well-tried and tested method is by calculating the heterozygosity index.
- Ecosystem diversity is most commonly measured in terms of the number and types of ecosystems in a defined area.
- Marine biodiversity can be measured functionally as well as structurally. Measurements of functional diversity can be measured simply by counting the number of functional groups or guilds in an ecosystem, or by identifying functional traits that can be used to calculate functional distances between species.
- Ways of avoiding unintended consequences of measuring biodiversity, including damage to the physical and living environment, are considered.

 Further reading

Bucklin, A., Steinke, D., and Blanco-Bercial, L. (2011) DNA barcoding of Marine Metazoa. *Annual Review of Marine Science*, 3, 471–508.

Available online at marine.annualreviews.org. This article's doi: 10.1146/ annurev-marine-120308–080950. Although a bit dated, a review which covers the essential features of eDNA.

Costello, M.J., Beard, K.H., Corlett, R.T., Cumming, G., Devictor, V., Loyola, R., Maas, B., Miller-Rushing, A.J., Pakeman, R., and Primack, R.B. (2016) Field work ethics in biological research: viewpoint of Biological Conservation editors. *Biological Conservation*, 203, 268–271.

A valuable perspective on biological ethics provided by highly respected marine scientists.

MarLIN, 2006. *BIOTIC - Biological Traits Information Catalogue.* Marine Life Information Network. Plymouth: Marine Biological Association of the United Kingdom. Available from <www.marlin.ac.uk/biotic>

A very useful resource for anyone studying benthic community ecology who wants to identify functional traits of marine organisms.

Petchey, O.L., and Gaston, K.L. (2006) Functional diversity: back to basics and looking forward. *Ecology Letters*, 9, 741–758. doi:10.1111/j.1461–0248. 2006.00924.x

An excellent account of functional diversity written by experts in biodiversity.

Thomson, S.A., Pyle, R.L., Ahyong, S.T., Alonso-Zarazaga, M., Ammirati, J., Araya, J.F., et al. (2018) Taxonomy based on science is necessary for global conservation. *PLoS Biology*, 16(3), e2005075. https://doi. org/10.1371/journal.pbio.2005075

An evocative and persuasive defence of the importance of taxonomy in conserving biodiversity.

 Discussion questions

5.1 Why is it necessary to refer to the species discovered during the COML as being only potentially new to science? And why are distribution records still being added to?

5.2 The COML project cost more than US$650 million by the time it was completed in 2010. Discuss whether or not this was money well spent.

5.3 The use of eDNA is emerging as a potentially valuable survey technique for aquatic environments and it is being promoted as a Citizen Science tool. Jeremy Biggs and his co-workers have already pioneered its use to study Great Crested Newts (*Triturus cristatus*) in freshwater habitats. Suggest a suitable Citizen Science project that would involve volunteers using eDNA to investigate an organism in a marine habitat.

To the Māori, the indigenous people of New Zealand, the reasons for conserving biodiversity are self-evident. Te moana, the coast and oceans of Aotearoa New Zealand, are central to the Māori identity. They recognize that their physical, cultural, and spiritual well-being are inextricably linked to the good health of their ecosystems. This is reflected in the Māori proverb 'Ko ahau te taiao, ko te taiao, ko ahau' ('I AM the environment and the environment IS ME').

Figure 6.1 Fishing in Whangaruru, Northland.

Photo: TS Images, Photo New Zealand

To many people whose lives are remote from the sea, the conservation of marine biodiversity is a costly business that can be justified only if it has tangible—and often, this means economic—benefits.

In the following sections we explore how marine biodiversity benefits us all both tangibly and intangibly.

Marine ecosystem services

The benefits of wild life and natural habitats are often described in terms of the ecosystems services they provide. 'Ecosystems services' is a broad concept covering all the benefits of nature to human wealth, health, and wellbeing. It's a concept used to make collaboration between scientists, decision-makers, and stakeholders easier when discussing environmental issues.

The ecosystems services concept is commonly categorized into four groups:

1. Supporting services: services that underpin all other ecosystem services by maintaining the conditions necessary for life. They include
   - the production of atmospheric oxygen;
   - nutrient cycling;
   - water cycling;
   - the creation and maintenance of habitats, e.g., coral reefs, mangrove swamps, and seagrass beds that act as spawning grounds and nurseries for a variety of species.

2. Provisioning services: services that are of direct benefit to human populations. Unlike the other ecosystem services, provisioning services are commonly traded and have obvious economic value. They include
   - aquaculture and fisheries, e.g., products used for food, fuel and fibre;
   - medical and pharmaceutical products derived from marine organisms, e.g., Green Fluorescent Protein from the jellyfish *Aequorea victoria*.

3. Regulating services: ecosystem processes with a regulatory role. They include the regulation of
   - seawater quality, e.g., a high marine biodiversity, which includes a variety of fish and zooplankton, moderates the effects of eutrophication by regulating the growth of phytoplankton and preventing the development of harmful algal blooms;

- air quality, e.g., removal of carbon dioxide from the atmosphere by marine biological processes such as photosynthesis;

- coastal erosion, e.g., coral reefs, mangroves, and seagrasses help protect coastlines from the erosive effects of storm damage;

- climate, e.g., phytoplankton emit dimethylsulfide (DMS) which contributes to the formation of clouds and blocks solar radiation, reflecting it back into space, moderating the effects of climate change; many scientists believe DMS production will be accelerated by global climate change, having a cooling effect which could help offset global warming; also many marine organisms play a part in storing and sequestering greenhouse gases;

- pests and diseases, e.g., the biological control of pests in the natural environment by cleaner fish in coral reefs, and by wrasses in aquaculture installations (see Figure 6.2).

Figure 6.2 *Gnathia* sp. are crustaceans belonging to the order Isopoda. The larval stage (see right below) is a blood-sucking ectoparasite of fish—this one is engorged with partly digested fish blood (hence the brownish colour). It is a common pest in aquaculture installations and aquaria, where it may be controlled by fish such as wrasse. *Gnathia* larvae have been called 'mosquitoes of the sea'.

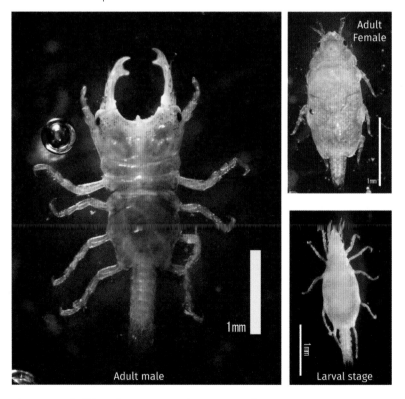

Photo: Y-zo/Wikimedia Commons (CC BY-SA 3.0)

- Marine organisms also provide a regulatory service by helping us monitor the marine environment (see Case study 6.3).

4. Cultural services: the non-material benefits that people obtain from ecosystems through spiritual enrichment, cognitive development, reflection, recreation, and aesthetic experience. They include contributions to

- symbolic and aesthetic values, e.g., marine life provides resources and inspiration for art in many parts of the world, and marine ecosystems have great spiritual and cultural significance;

- recreation and tourism, e.g., the marine environment and its biodiversity contribute to many kinds of recreational activities such as surfing, and ecotourism activities (e.g., whale watching) that have enormous economic benefits;

- cognitive services, e.g., the role that the sea and marine life have in maintaining mental as well as physical health. Cognitive services also include the contribution that marine biodiversity makes to our education and scientific knowledge (see Scientific approach 6.1).

The ecosystem concept has been criticized for focusing mainly on economic values, that it is too anthropocentric, and that it promotes an exploitative human–nature relationship (see the review article by Matthias Schröter and his co-authors in Further reading). The Māori have adapted the concept so that it reflects the importance they give to cultural values.

The Māori use their own ecosystem framework to make management decisions about the marine environment. It incorporates the ecosystem concept, but 'cultural values' underpin all aspects of the framework (Figure 6.3). For Māori, 'cultural values' include things that are tangible or intangible, material or non-material, of use or no use, qualitative or quantitative.

Oxygen provision

Most people associate atmospheric oxygen with the process of photosynthesis. But there are probably few who realize that there are two types of photosynthesis: anoxygenic and oxygenic. Only oxygenic photosynthesis generates oxygen, as a by-product of solar radiation splitting water to release electrons which act as the initial donors in non-cyclic photophosphorylation. In anoxygenic photosynthesis, water is not split. A different molecule, such as hydrogen sulfide, is used as an electron donor for non-cyclic photophosphorylation; no oxygen is produced.

Anoxygenic photosynthesis is the type of photosynthesis that occurred about a billion years before oxygenic photosynthesis became common; it is still carried out today by several groups of bacteria. There is much speculation about the evolution of oxygenic photosynthesis, but most scientists think it originated in an ancestor of cyanobacteria (blue-green algae) when an anoxygenic photosystem gave rise to a water-splitting photosystem.

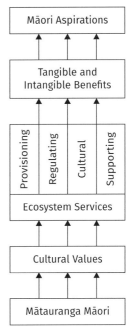

Figure 6.3 Cultural values underpin all aspects of the Māori ecosystem framework. Mātauranga Māori refers to Māori knowledge and philosophy.

Source: From Harmsworth, G.R., and Awatere, S. (2013). Indigenous Māori knowledge and perspectives of ecosystems. In Dymond, J.R. ed. Ecosystem services in New Zealand – conditions and trends. Manaaki Whenua Press, Lincoln, New Zealand. pp. 274–286

Whatever the first organism to perform oxygenic photosynthesis was, cyanobacteria are probably the most important providers of atmospheric oxygen today. Especially important is one genus of cyanobacteria called *Prochlorococcus* (see Figure 6.4).

These blue-green algae were first isolated in 1986 from the Sargasso Sea, by oceanographer Sallie (Penny) W. Chisholm of the Massachusetts Institute of Technology. In 2019 she was awarded the $750,000 Crafoord Prize for her discovery and subsequent studies of cyanobacteria.

Prochlorococcus has been described as '... *just a green mote, floating in vast numbers in the world's oceans'*, but the research of Penny Chisholm and others reveals that *Prochlorococcus* is probably the most abundant photosynthetic organism in the marine environment. *Prochlorococcus* has many ecotypes—distinct types found in particular habitats. It is estimated that collectively the various ecotypes use about 80 000 genes. This amazingly high genetic diversity enables *Prochlorococcus* to be ubiquitous in the ocean between the latitudes of 40° N and 40° S latitude, and to thrive in waters throughout the euphotic zone from the sunlit surface to the dim depths at 200 metres.

The algae are so small that it would take 10 of them to fit across the width of a red blood cell. But what they lack in size, they gain in numbers. In certain locations, the number of *Prochlorococcus* cells exceeds 100 000 per millilitre of seawater.

Collectively, the many groups of blue-green algae, phytoplankton, and seaweed generate about half of all the oxygen in the Earth's atmosphere. One answer to the question 'What can marine biodiversity do for us?' is that it provides approximately 50% of the oxygen in every breath that we take.

Figure 6.4 A false-coloured electron micrograph of *Prochlorococcus marinus*, a small (each cell is about 0.6 μm wide) but globally significant producer of atmospheric oxygen.

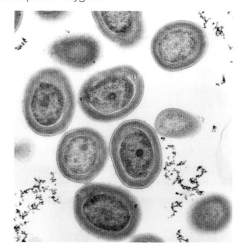

Source: Image taken by Luke Thompson from Chisholm Lab and Nikki Watson from Whitehead, MIT, 2007. The Chisholm Lab have given permission to use this image under the Creative Commons CC0 1.0 Universal Public Domain Dedication.

Carbon sequestration and climate change mitigation

Carbon sequestration, the process by which carbon dioxide is removed from the atmosphere, is one of the most important regulating services carried out by marine organisms. It has moderated the build-up in the atmosphere of carbon dioxide (produced by natural events such as volcanoes and also the primary greenhouse gas emitted by human activities), thereby reducing climate change. According to a 2019 United Nations report, climate change is '... *the defining issue of our time and it is happening even more quickly than we feared*'.

Climate change is a serious threat to human welfare now and will continue to be a threat for future generations. In 2014 Rajendra K. Pachauri, as Chairperson of the Intergovernmental Panel on Climate Change (IPCC), stated that nobody on the planet will be untouched by climate change.

Marine biodiversity makes a major contribution to climate change mitigation. In coastal wetlands, mangroves, salt marshes, and seagrasses sequester and store huge amounts of organic carbon. Marine and maritime vegetation reduce the erosion of underlying muddy sediments that are rich in organic matter. Without vegetation cover, the mud is easily disturbed by water movements, and releases carbon dioxide back into the atmosphere.

Phytoplankton, algae, and marine plants extract carbon dioxide from the atmosphere for photosynthesis, and the carbon is incorporated into their bodies. Once fixed by photosynthesis, carbon becomes part of the marine food web. The fate of this carbon has two main outcomes: it may be respired and released back to the ocean and potentially to the atmosphere as carbon dioxide, or the carbon may be fixed as Particulate Organic Carbon (POC: dead bodies, faeces, and partly decayed matter). Eventually, POC finds its way down to marine sediments in the deep ocean where it may be locked away for centuries (see Scientific approach 1.3). The formation of POC and the subsequent transfer of carbon to the deep ocean through gravitational sinking is the major part of a carbon sequestering system called the biological carbon pump (BCP). The BCP is responsible for exporting an estimated 0.6 to 1.3% of marine primary production into sediments at ocean depths over 2000 metres.

Another carbon sequestering system called the microbial carbon pump (MCP) was discovered in 2010. The MCP does not necessarily involve physical transportation. It depends instead on the microbial loop in which bacteria chemically transform dissolved organic matter (DOM) from rapidly degradable forms of carbon to slowly degradable forms. It's estimated that the MCP is responsible for storing about 0.4% of the annual marine primary production. BCP and MCP act simultaneously, but the balance between them is thought to depend on nutrient levels in the marine environment.

The BCP and MCP play vital roles in sequestering carbon dioxide from the atmosphere to the oceans and help maintain atmospheric carbon dioxide at levels significantly lower than would be the case if they did not exist.

Carbon sequestration is not only affected by the activities of marine autotrophs and microbes, it is also affected by organisms that fix carbon as calcium carbonate in their shells and skeletons. These organisms span the whole phyletic range from microscopic single-celled plankton to the Blue whale. As well as fixing carbon into skeletal materials, vertebrates facilitate carbon sequestration in a number of other ways (see Case study 6.1).

Case study 6.1
Whale poo and other carbon matters

In a 2014 report entitled *Fish Carbon*, Steven Lutz and Angela Martin highlight seven biological mechanisms provided by marine vertebrates that result in carbon sequestration, and one mechanism that moderates acidification (Figure A). Lutz and Martin refer to these carbon sequestration methods as:

- **Trophic Cascade Carbon**, in which food web dynamics help maintain the carbon storage and sequestration function of coastal marine ecosystems (for example predation and herbivory in seagrass meadows and kelp forests—see Case study 4.1);

- **Biomixing Carbon**, resulting from the turbulence and drag associated with the movement of marine vertebrates mixing nutrient-rich deep

Figure A Marine vertebrate carbon services: eight carbon pathways, pumps, and trophic cascades.

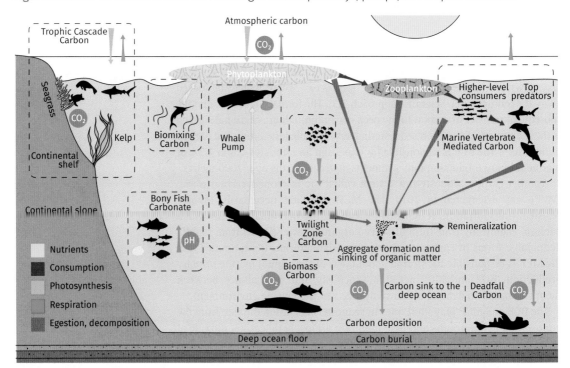

Redrawn from figure jointly produced by GRID-Arendal and Blue Climate Solutions

water with water towards the sea surface; this enhances primary production and increases the uptake of dissolved CO_2;

- the **Whale Pump**, in which nutrients from whale faeces (poo) stimulate primary production by phytoplankton and increase the uptake of dissolved CO_2;

- **Twilight Zone Carbon**, a form of carbon sequestration caused by mesopelagic fish feeding in the upper ocean layers during the night and transporting consumed organic carbon to deeper waters during daylight hours;

- **Biomass Carbon**, the storage of carbon as biomass in marine vertebrates; the carbon is retained in the ocean throughout the natural lifetimes of the vertebrates; larger individuals with long life spans store proportionally greater amounts of carbon over prolonged timescales;

- **Deadfall Carbon**, in which the carcasses of large pelagic marine vertebrates sink through the water column, exporting carbon to the ocean floor where it becomes incorporated into the benthic food web, and is sometimes buried in sediments (see 'A closer look at whale fall' in Chapter 1, and Scientific approach 1.3);

- **Marine Vertebrate Mediated Carbon**, which refers to carbon sequestration by marine vertebrates that consume and repackage organic carbon through marine food webs, and transport the carbon in faecal material that sinks to the sea floor.

In a 2020 article in *The Niche* (a magazine of the British Ecological Society) Dr Emma Cavan of Imperial College refers to krill as the 'Unsung Ocean Superhero'. In Chapter 2, we looked at krill as key components of Antarctic food webs. Emma's recent research has shown that they are much more than that. She writes, '*Antarctic krill do not just feed whales and penguins, they have other superpowers. They fertilize the ocean and help regulate both the ocean's carbon and the air we breathe.*' She supports this claim with a figure that shows the role of Antarctic krill in biogeochemical cycles (see Figure B).

- Krill feed on algae releasing nutrients (such as iron and ammonium) through faeces and excreta, further stimulating algal growth and CO_2 drawdown.

- Krill faecal pellets sink to the deep sea, with some consumed by animals and bacteria as they sink.

- Krill swimming from the deep bring nutrient-rich water to the surface stimulating algal growth.

- Young krill live under sea ice and transfer carbon deeper in the ocean than adult krill.

- Consumption of krill by whales stores krill carbon in the whale for decades until the whale dies and sinks to the seafloor.

Figure B The role of Antarctic krill in biogeochemical cycles.

Source: Cavan, E.L., Belcher, A., Atkinson, A. et al. The importance of Antarctic krill in biogeochemical cycles. Nature Communications, 10, 4742 (2019).

❓ Pause for thought

Stephen Lutz and Angela Martin focus on the role of whales in sequestering carbon (Figure A). Emma Cavan looks at carbon sequestration from an Antarctic krill perspective (Figure B). Which do you think plays the most significant role in carbon sequestration, whales or krill - and why?

Marine model organisms

A model organism is a species with special characteristics that make it suitable for extensive studies into fundamental questions about biological structures and processes.

Researchers use model organisms in the expectation that any discoveries they make will provide insights into the workings of other species, including our own. The criteria used to identify suitable marine model organisms include availability, robustness, ease of maintenance and breeding under laboratory conditions, and evolutionary, ecological, and economic significance.

One reason for the frequent choice of marine organisms as models for biomedical research is because more phyla are found in the ocean than anywhere else. Phyla are significant in this context, because each phylum is characterized by a distinctive body plan. There is no definitive list of phyla, but it is generally agreed that the higher diversity of major life types in the sea offers more possibilities for using marine organisms as models to explore various biological processes.

Echinoderms (starfish, sea urchins, and brittlestars) are, like us, deuterostomes. They share a common origin with mammals (you can find out more about this in another title in this series, *Animal developmental biology: embryos, evolution, and ageing*), and have been particularly useful marine model organisms. For example, in 1882, Elie Metchnikoff (1845–1916) carried out a key experiment on starfish which led to it becoming an important marine model organism. He punctured a starfish larva (see Figure 6.5) with the thorn of a rose. On the following day, he observed that small, motile cells had surrounded the thorn and appeared to be devouring the foreign body in a process called phagocytosis. After making similar discoveries in a range of other organisms, Metchnikoff suggested that phagocytosis might be a fundamental defensive mechanism which is widespread in the animal kingdom, writing, '*There is only one constant element in immunity, whether innate or acquired, and that is phagocytosis.*'

Figure 6.5 Starfish larvae like this are good model organisms for directly observing the internal processes of a living animal—they are almost transparent, and are easy to breed and maintain in the laboratory.

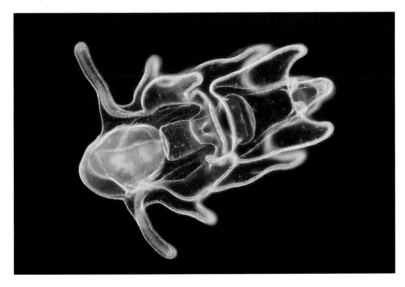

Photo: Wim van Egmond/Science Photo Library

Metchnikoff's work on starfish laid the foundation for further research on cellular and comparative immunology, which in turn has shed much light on the role of phagocytosis in fighting infection and disease in humans.

Other research using marine model organisms has shed light on many biological structures and functions at all levels of organization, from the subcellular to the whole organism and beyond. For example, marine model organisms have been used to discover:

- cellular, biochemical, and gene regulatory mechanisms controlling fertilization. sea urchin embryos have been a major experimental model in this work for over a century;

- nerve cell transmission: the discovery of the mechanisms of the action potential was made using giant axons of squid, *Loligo pealei*, at the Laboratory of the Marine Biological Association in Plymouth by Sir Alan Hodgkin and Sir Andrew Huxley. They were awarded the Nobel Prize for Physiology or Medicine in 1963 for this ground-breaking work, which laid the foundations for much of our current understanding of how the nervous system functions;

- nerve-to-nerve synaptic transmission: studied in the electric organs of marine rays (see Figure 3.8), which have a synapse density more than 100 times that in human muscle. This made possible the purification of a large variety of the synaptic molecules including the receptor for acetylcholine, a major neurotransmitter in the nervous system of all animals;

- immune systems that control bacteria and fungi: not only starfish, but also sea urchins, tunicates, and sharks have been used to improve our understanding of how the body fights infections and disease;

- cellular and molecular mechanisms responsible for learning and memory: the marine snail *Aplysia* is widely used for investigating the cellular regulation of behaviour, including learning and memory; see Case study 6.2.

Case study 6.2
A Nobel Prize ... thanks to the memory of sea hares

Aplysia californica, commonly called a sea hare, is one of the most celebrated marine model organisms. It gained its status by being the experimental animal used by Professor Eric Kandel for his Nobel Prize-winning work on signal transduction in the nervous system. This is a key mechanism in learning and memory in all animals. It takes place at synapses, the specialized points where nerve cells communicate with each other. Kandel and his team of

Figure A *Aplysia californica*, a species of sea slug that feeds on seaweeds. It can grow up to 75 cm in length with a mass of nearly 7 kg.

Photo: Chad King / NOAA MBNMS

co-workers found that synapses involved in simple reflexes can be modified by learning.

Eric Kandel has a rich and varied academic background. He graduated at Harvard with a history degree, then changed direction and completed a medical degree with the intention of becoming a psychoanalyst. But during his medical studies, he began to think about problems of psychiatry and psychoanalysis in biological terms. His interest was mainly in human psychology, partly motivated by his desire to understand why people living in a highly cultured country can do unspeakable things—Kandel is Jewish; born in Vienna in 1929, he fled from Austria to the USA in 1939 to escape Nazi persecution.

After qualifying as a medical doctor, despite his interest in human psychology, Kandel turned his attention to invertebrate subjects. He was greatly influenced by neurobiologists who demonstrated that tissue preparations obtained from invertebrates could be used to gain new insights into the functioning of nervous systems in all animals, including ourselves.

Stephen Kuffler was one of the neurobiologists whose work particularly impressed Kandel. Kuffler used crayfish and lobsters in experiments which demonstrated that sensory neurones in invertebrates were, structurally and functionally, similar to those in humans. He attributed his success to choosing a species in which the nervous tissue under investigation was accessible, easily isolated, and easy to see.

After receiving his Nobel Prize, Kandel reportedly said that he had learned from Kuffler the importance of having an experimental preparation suitable to testing the questions he wanted to answer. Inspired by Kuffler, Kandel decided to adopt a **reductionist** approach to his research, and use nervous tissue from an invertebrate to study learning and memory. A reductionist approach allows the highly specific circuitry controlling a behaviour to be identified, defined, and examined to see if there are any changes in the circuitry following learning.

Undeterred by those who thought a reductionist approach was inappropriate, Kandel searched for an experimental animal that exhibited a simple behaviour, modifiable by learning. His ideal organism was one with an easily observed behaviour controlled by a small number of large and accessible nerve cells, so he could relate the animal's observed behaviour to events occurring in the cells controlling it.

Kandel found his ideal organism in *Aplysia californica*.

- Its nervous system is relatively simple (it has only about 20 000 nerve cells, compared with about 100 billion found in the human brain);
- many *Aplysia* nerve cell bodies are distinctive and relatively large (up to 1 mm for the largest neurones);
- the large neurones are found in the same locations in the brains of all individuals;
- *Aplysia* has a simple form of behaviour which can be readily modified by two types of learning: habituation and sensitization;
- *Aplysia* is attractive to look at as well as to work with, according to Kandel! This is not a bad reason for choosing an animal that you are planning to spend many years working with.

Aplysia exhibits a piece of readily elicited and quantified behaviour. A gentle touch of the siphon (see Figure B) triggers the gill reflex, resulting in the gill being pulled in under the mantle. The neuronal pathway for this reflex action is wonderful, as far as an experimenter is concerned, in its simplicity: it consists of a single sensory neurone connected, via a synapse, to a single motor neurone.

Repeated tactile stimulation results in habituation (a decrease in responsiveness) of the withdrawal reflex. Short-term habituation, typically measured in hours or days, involves molecular changes in pre-synaptic connections. Long-term habituation, in which a reduction in responsiveness is retained for more than 1 week, seems to be linked to a progressive decrease in post-synaptic excitatory potentials.

For a sea slug to habituate to a tactile stimulus, its nervous system must be able to register and store the information that a previous stimulus has occurred. Therefore, studies in habituation can reveal how memory (storage and retention of information) is achieved: the interval of time between stimuli that results in habituation taking place can provide a measure of memory retention. In recent years, age-related changes in habituation in *Aplysia* have been used to study age-related changes in memory retention in humans.

Sensitization is a slightly more complex form of learning. It has been described as the mirror image of habituation, as it is characterized as an enhancement of a reflex response to one stimulus as a result of the presentation of another, noxious, stimulus. Whereas in habituation, repeated stimulation depresses responsiveness due to a decrease in neurotransmitter molecules released by the sensory neurones, a sensitizing stimulus enhances

Figure B A simple learned behaviour taken from Kandel's own work (**A**) *Aplysia* dorsal view with the mantle shelf retracted for a better view of the gill. If the siphon is touched gently, it contracts and the gill withdraws. If you give an unpleasant, sensitizing stimulus to another part of the body of *Aplysia*, such as the tail, the strength of the withdrawal reflex of both the siphon and the gill increases. (**B**) The graph demonstrates how *Aplysia* develops long-term memory when a stimulus is repeated. Before training, a gentle touch to the siphon causes a weak, brief siphon and gill withdrawal reflex. After a single unpleasant, sensitizing, shock to the tail, a gentle touch produces a much larger siphon and gill reflex withdrawal response. This increased response lasts around an hour after the unpleasant tail stimulus. If the tail shocks are repeated, the size of the gill response increases and the effect is remembered for longer.

A

Gill withdrawal reflex   **Sensitization**

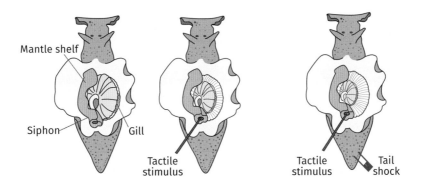

B

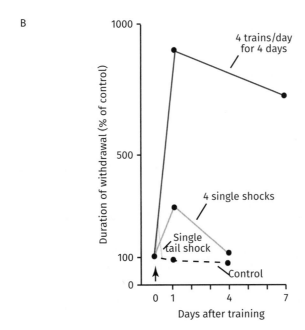

responsiveness due to an increase in the amount of transmitter released per impulse.

The discoveries of Kandel and others working with *Aplysia* have shown that a simple marine model organism can provide amazing insights into the workings of the nervous system in all animals, including humans. Researchers, building on the work of Kandel, continue to use *Aplysia californica* and related species to study learning and memory, and to explore the causes of neurological and psychological diseases.

❓ Pause for thought

Kandel's reductionist approach is traditional in western science. In neurobiology, it is exemplified by Hodgkin and Huxley's work on the giant axon of squid, as well as Kuffler's work on crustacea. However in the 1960s, many eminent neuroscientists were reluctant to adopt a reductionist approach to the study of behaviour, especially human behaviour. Discuss possible reasons for this.

Wealth creation

In 2006, Defra (the Department for Environment, Food and Rural Affairs) produced a report entitled *Marine Biodiversity: an economic valuation*. The report gave a valuation of the goods and services provided by marine biodiversity; it was produced to provide evidence for legislation designed to protect marine ecosystems and biodiversity in UK waters. The authors of the report adopted a Total Economic Valuation (TEV) approach which ensured the valuations included all economic benefits resulting from goods and services linked to marine biodiversity in the UK. Table 6.1 gives an overview of these valuations.

The authors of the Defra report admit that the **TEV** approach is controversial, and recognized its shortfalls. One shortfall is that the authors were limited to data available at the time. But the authors point out that the 'Valuation data not available' comment in their table does not imply that a good or service is without value, only that the information is not available.

The authors stress that their valuations are only indicative estimates. This is even truer today: 2020 valuations would be much higher. However, to paraphrase the authors, the strength of valuing marine biodiversity in monetary terms lies less in the absolute values given, but more in its capacity to raise awareness of the importance of marine biodiversity.

The 'guesstimates' given in the table relate to the monetary value of marine biodiversity in the UK, in the early 2000s. Today, global marine diversity is valued in trillions of dollars. According to a 2017 FAO report,

Table 6.1 An overview of the monetary value of goods and services provided by UK marine biodiversity

Good/service	Definition	Monetary value per annum, UK £, 2004	Link to biodiversity low (1) to high (5)
Food provision	Plants and animals taken from the marine environment for human consumption	£513 million	3
Raw materials	The extraction of marine organisms for all purposes, except human consumption	£81.5 million	3
Leisure and recreation	The refreshment and stimulation of the human body and mind through the perusal of, and engagement with, living marine organisms in their natural environment	£11.7 billion	3
Resilience and resistance	The extent to which ecosystems can absorb recurrent natural and human perturbations and continue to regenerate without slowly degrading or unexpectedly flipping to alternate states	Valuation data not available	5
Nutrient cycling	The storage, cycling, and maintenance of availability of nutrients mediated by living marine organisms	£800 – £2320 billion	4
Gas and climate regulation	The balance and maintenance of the chemical composition of the atmosphere and oceans by marine living organisms	£0.4 – £8.47 billion	5
Bioremediation of waste	Removal of pollutants through storage, dilution, transformation, and burial	Valuation data not available	5
Biologically mediated habitat	Habitat which is provided by living organisms	Valuation data not available	5
Disturbance prevention and alleviation	The dampening of environmental disturbances by biogenic structures	£0.3 billion plus £17– 32 billion capital costs	4
Cultural heritage and identity	The cultural value associated with the marine environment, e.g., for religion, folk lore, painting, cultural and spiritual traditions	Valuation data not available	3

(Continued)

Table 6.1 *(Continued)*

Good/service	Definition	Monetary value per annum, UK £, 2004	Link to biodiversity low (1) to high (5)
Cognitive values	Cognitive development, including education and research, resulting from marine organisms	£317 million	4
Option use value	Currently unknown potential future uses of the marine environment	Valuation data not available	5
Non-use values: bequest and existence	Value which we derive from marine organisms without using them	£0.5 –1.1 billion	5

Based on Table 1 in Beaumont, N., Townsend, M., Mangi, S., and Austen, M.C. (2006) *Marine Biodiversity: An economic valuation.* Defra, UK.

the global value of fish exports, just one component of the provisioning services provided by marine organisms, was worth more than US$152 billion annually.

A closer look at 'option use values'

In Table 6.1, the 'option use value' of marine biodiversity goods and services refers to the economic benefits of 'currently unknown potential future uses of the marine environment'. Marine bioprospecting involves systematically searching for interesting and unique genes, molecules, and organisms from the marine environment that might be useful to society and/or have potential commercial benefits. Like oil prospecting, bioprospecting is highly speculative and can be very costly. To minimize costs and maximize the chances of finding valuable bioproducts, marine bioprospectors gather as much evidence about potential uses as possible.

Among the most lucrative bioproducts are medically useful chemicals. Many are used to treat cancer, bacterial infections, and other diseases. Others, such as Green Fluorescent Proteins (GFPs) are used in diagnostic procedures and in medical research.

Green fluorescent protein from the jellyfish *Aequorea victoria* is one of the most highly prized chemicals in medicine. GFP radiates a visible green fluorescent light when exposed to light in the blue to ultraviolet part of the spectrum. When attached to proteins otherwise undetectable under the microscope, GFP has, literally, shed light on many processes which take place within cells and whole organisms. GFPs attached to pathogens have been used to track the colonization, proliferation, tenacity, and spread of the pathogens in live animals. GFPs are also used widely in cancer research to label and track cancer cells.

Figure 6.6 A Snakelocks anemone (*Anemonia sulcata*) contains symbiotic algae (zooxanthellae like those in coral). The deep green colour of the tentacles is due to the presence of chemicals homologous to the Green Fluorescent Protein of *Aequorea victoria*.

Photo: © Michael Kent

GFPs occur in many other coelenterates, including the Snakelocks anemone (Figure 6.6), but they all differ slightly. They have been isolated and screened for their bioactivity to assess their potential use.

In the ongoing search for new, more effective and medically useful bioactive chemicals, marine bioprospectors are turning their attention to some of the least explored marine environments. One such environment is the Arctic Sea where bioprospectors at the Marbio Analytical Laboratory of the Arctic University of Norway are screening cold-adapted organisms for useful natural marine products.

Cold-adapted Arctic organisms include fish, invertebrates, and microorganisms such as microalgae, bacteria, and fungi. Screening uses techniques such as Liquid Chromatography–Mass Spectrometry (LC-MS) to analyse complex mixtures of biochemicals derived from the organisms. LC-MS enables new chemicals to be discovered, and potentially active ingredients to be identified.

Newly discovered chemicals are then isolated and tested for their bioactivity.

Recent work by the Marbio researchers includes screening chemicals from cultures of *Pseudomonas* (a bacterium with strains that cause lung disease) isolated from the Arctic Sea; so far, several new types of antibacterial and cytotoxic drugs have been discovered, which are being tested for clinical use.

Many of the services that marine biodiversity will provide in the future will not be the result of systematically bioprospecting for something useful.

They will evolve from research driven by all sorts of factors, not least an insatiable curiosity about the natural world around us (see Scientific approach 6.1).

Scientific approach 6.1
A marine biologist's motivation

How science works depends not only on the methods and techniques used, but also on the motivation to do it. In 1974, when Rodney Phillips Dales became a Professor at the University of London, he presented a lecture titled In Praise of Zoology in which he reflected on the relative merits of **pure research** and **applied research**. He talked about his research findings on the pigments in *Neoamphitrite*, its links to medical research on **porphyria**, and his motivation for carrying out the research. He said '... *I could not know where the work would lead, or whether the results would be important or not. Nothing would delight me more if some of these facts proved in the event to lead to some medical advance, but this was not the motive for the work.*' *Neoamphitrite figulus* is an especially plump, reddish brown polychaete worm that grows up to 25 centimetres in length. It feeds by extending long ciliated tentacles over the muddy surface of the substrate in which it lives.

N. figulus was Professor Dales's favourite research organism. His initial interest was sparked by noticing that individuals varied in colour: some were brown and others red. Curious to find out what caused the colour

Figure A *Neoamphitrite figulus*, a polychaete worm which became the favourite research organism of Professor Rodney Phillips Dales.

Photo: Nature Photographers Ltd / Alamy Stock Photo

variation, he carried out detailed biochemical analyses. These showed that red individuals contained carotenoids and haemoglobin, whereas brown individuals contained these two pigments and another called coprahaematin III, a pigment that had never been found in living organisms before. The formation of coprahaematin III was due to prolonged exposure to light. Consequently, individuals became more brown with age. A literature search revealed, unexpectedly, that the formation of coprahaematin III was similar to that of pigments associated with the inherited human disease porphyria.

Professor Dales argued that much of the research that leads to medical advances starts without its applications in mind. Some readers might be surprised that he discussed the unravelling of the structure of DNA as an example of pure research. However, it is only with the benefit of 2020 hindsight that we can see how Watson and Crick's discoveries of the double helix led to an amazing range of uses of DNA in the life sciences and beyond.

When Watson and Crick started their research, they had little idea how their findings would be used. Even in 1974, 20 years or so after the discovery of the DNA double helix, there was little indication that their discovery would be of much practical use.

Reflecting the scientific view of the time, Professor Dales said in his 1974 lecture that, '*No one would deny that Watson and Crick's modest note in* Nature, *published twenty years ago, in which they suggested a double helical structure for DNA, has had momentous results. Yet MacFarlane Burnet [a Nobel Prize-winning biologist] has observed that it has made no significant contribution to medicine and is convinced that it never will. But the effect on biology and our thinking generally has been immense.*'

Read James Watson's fabulously feisty inside story of how the DNA double helix structure was discovered, and you will get a good idea about what motivated him and Crick to study the structure of DNA (see Further reading). It was a complex mix that included the search for fame and recognition, as

Figure B James Watson and Francis Crick examining their model of DNA. This revolutionized our understanding of the fundamentals of all aspects of biology, including the nature of marine biodiversity.

Photo: A. Barrington Brown/Science Photo Library

well as the pursuit of scientific knowledge for its own sake. Competing to be the first research team to discover the chemical and physical nature of inheritable material was a key part of their motivation. They had enough to occupy their minds without having to think how their results would be applied after they made their discovery.

In his lecture, Professor Dales gave many examples of how 'pure research', or 'basic science', conducted in the pursuit of fundamental biological truths, has led to results with applications of immense value. He also gives examples of how some applied, mission-, or goal-directed research has brought us closer to understanding those fundamental truths. He emphasizes that the main distinction between these two approaches to science lies not in the work itself, but in the motives for the work. The primary motivation of Professor Dales and many other marine biologists is simply the pursuit of knowledge, wherever it might lead. However, the desire to achieve something that might be useful is also a strong motivator in biological research … especially for those funding the research.

❓ Pause for thought

Should scientific work with no apparent practical value receive similar levels of funding as research with the specific aim of solving particular human problems? Discuss, looking for evidence to support your ideas.

Keeping us healthy

Many of the goods and services provided by marine organisms affect human health in an obvious and direct way. Oily fish, such as mackerel and tuna, are rich in omega fatty acids, vitamin D, and high quality protein that help to keep us nutritionally healthy. Seaweeds, such as *Porphyra* sp. (Figure 6.7), commonly regarded as no more than slimy and slippery things, are packed with minerals and vitamins with significant health benefits. *Porphyra* is used to make Welsh laver bread, and Japanese Nori rice balls and rolls (nori is the Japanese word for edible seaweed). There are over 70 species of *Porphyra* worldwide; each species has a slightly different taste.

A wide variety of marine organisms act as ecosystem sentinels. A sentinel is a relatively large, conspicuous organism in an ecosystem, often a top predator. Observing their timely and measurable responses to environmental changes warns us of potential or actual harm to the health of the whole ecosystem (see Case study 6.3).

As well as keeping us physically healthy, marine organisms in their seemingly infinite variety help to keep us mentally healthy. Although it's sufficient for most of us to 'know in our bones' that being close to the sea and its creatures is good for us, there's much scientific evidence to support the notion. In an article in *The Marine Biologist* (see Further reading),

Figure 6.7 Purple laver (*Porphyra* sp.) is a cold-water, edible, translucent seaweed renowned for being a good source of protein, vitamins, and minerals, including vitamins B, C, E, and beta-carotene.

Photo: © Michael Kent

Michael Depledge, Ben Wheeler, and Mat White wrote, '*It is notable that in the UK the National Health, primarily focused on treating rather than preventing disease, costs the taxpayer ca. £110 billion pounds per year. Fostering access to marine environments that promote well-being could contribute significantly to reducing these costs.*'

Case study 6.3
The sperm whale, an extraordinary ecosystem sentinel

There are two main types of ecosystem sentinel: **elucidating sentinels** indicate past or ongoing changes in the ecosystem; **leading sentinels** help scientists predict a change in the marine environment.

The sperm whale ticks most of the boxes for characteristics which make it an extraordinarily good elucidating sentinel of the global marine environment: it is conspicuous, accessible, and observable, and it can provide ecosystem information across large spatiotemporal scales, enabling it to reveal components in the marine environment which would otherwise go unobserved. Because of its exceptionally wide distribution, the sperm whale is considered a worldwide sentinel of ocean pollution and health.

Several studies of marine pollutants, including DDT, Persistent Organic Pollutants (POPs), and lead, have involved observations of sperm whales. Researchers take advantage of the whale's wide distribution and its status as top predator (it feeds on a range of fish and other organisms, most notably giant squid). The bioaccumulation of pollutants in sperm whale tissue make them good sentinels of the health of the world's oceans. One technique used in marine pollution research involves an arrow being shot into the surface flesh of a whale, so the tip contains a small core of skin and flesh which is removed for analysis. With precise shooting and hygiene precautions, whales are not harmed any more than they are by the seabirds that peck at their skin—in the subantarctic waters off South Georgia, researchers have recorded giant petrels landing on the backs of sperm whales to obtain a quick snack of skin and blubber.

In 2015, Laura Savery and her collaborators established a global baseline of oceanic lead (Pb) concentrations using free-ranging sperm whales as an ecosystem sentinel. Lead is an oceanic pollutant of global concern. Skin biopsies were collected during the voyage of the *Odyssey*, a vessel from which researchers obtained biopsies from sperm whale in 17 oceanic regions. The results provided the first global toxicological dataset for lead in cetaceans; they confirmed that lead is widely distributed, with particularly high concentrations of the pollutant in some regions.

Figure A A sperm whale, *Physeter macrocephalus*. It can grow up to 16 metres in length and weigh more than 40 tonnes.

Photo: Martin Prochazkacz/Shutterstock

⦿ Pause for thought

- Why is sperm whale tissue particularly effective in accumulating pollutants such as lead, DDT, and POPs?
- Suggest how modern technology, such as that associated with the Global Positioning System, might enable sperm whale to be used as an ecosystem sentinel of the marine environment without needing to take flesh biopsies.

⦿ Chapter summary

- The concept of ecosystem services is used to consider what marine biodiversity does for us.
- Marine ecosystem services are categorized into four types: supporting services, provisioning services, regulating services, and cultural services. Examples are described for each type.
- The release of oxygen as a by-product of photosynthesis by marine primary producers is described and its importance discussed.
- The sequestration of carbon by marine organisms is considered in relation to climate change.
- Examples of the use of marine model organisms in research to gain insights into the functioning of our own and other species are described.
- The monetary values of goods and services provided by marine biodiversity are given as tangible examples of what marine biodiversity can do for us.
- Medical products derived from marine organisms show how marine biodiversity benefits physical health.
- Marine ecosystem sentinels are described, with sperm whale given as a prime example of a marine organism that can be used to indicate past and ongoing changes in the global marine ecosystem that might affect human health.
- The benefits of marine biodiversity to human mental health are considered.

⦿ Further reading

Depledge, M., Wheeler, B., and White M. (2014) Seas, society, health and wellbeing. *The Marine Biologist*, Issue 3 from the MBA website at www.mba.ac.uk.
A brief and clear introduction to the importance of coastal ecosystems to health and well-being.

Harmsworth, G.R., and Awatere, S. (2013) Indigenous Māori knowledge and perspectives of ecosystems. In Dymond, J.R., ed., *Ecosystem services in New Zealand—conditions and trends*. Available as a pdf from http://www.mwpress.co.nz/__data/assets/pdf_file/0007/77047/2_1_Harmsworth.pdf.
A fascinating paper describing the ecosystem framework that the Māori use to make ecosystem management decisions.

Hazen, E.L., Abrahms, B., et al. (2019) Marine top predators as climate and ecosystem sentinels. *Frontiers in Ecology and the Environment*, 17(10), 565–574. doi:10.1002
A multi-author review discussing how top predators can reveal how marine ecosystems function, identify hidden risks to human health, and predict future change.

Schroter, M., van der Zanden, E.H., et al. (2014) Ecosystem Services as a Contested Concept: A Synthesis of Critique and Counter-Arguments. *Conservation Letters*, November/December 2014, 7(6), 514–523.
An insight into how a widely accepted scientific concept can generate opposition as well as support.

Watson, J. (2010) *The double helix: a personal account of the discovery of the structure of DNA*. W&N reprint edition.
Probably one of the most revealing accounts of world-class science in action; it shows how science can be very competitive. Best read alongside *Rosalind Franklin: The Dark Lady of DNA*, by Brenda Maddox.

 Discussion question

6.1 With friends, family, or fellow students interested in marine wildlife, discuss what marine biodiversity does for you personally and for you as a member of society.

7 WHAT ARE WE DOING TO MARINE BIODIVERSITY?

Soft-bodied and with no mouth or razor-sharp teeth, a box jellyfish (Figure 7.1) looks pretty innocuous. But it is probably the most feared marine creature in Australian waters: even champion surfers quake at the sight of its bright body and trailing tentacles. This beautiful box-shaped beast is responsible for more human deaths in Australian waters than saltwater crocodiles, sea snakes, and sharks combined. A single box jellyfish has enough venom to kill more than 50 people. What makes this coelenterate particularly dangerous is that its venom is fast-acting: it can take less than five minutes from being stung to being dead. Those stung feel a searing heat as well as intense pain; they often think that applying ice might help. But research has shown that jellyfish stings are highly heat sensitive; immersion in hot water leads to better outcomes than ice.

Students visiting the Natural History Museum in Kensington are often asked to search the museum for the exhibit labelled 'The most dangerous animal in the world'. Students encounter many contenders for the title, including the great white shark, pufferfish, box jellyfish, and a deep-sea giant squid, but none of these, not even the box jellyfish, has the label they are looking for.

Figure 7.1 The sea wasp, *Chironex fleckeri*, is probably the deadliest of the 50 or so species of box jellyfish.

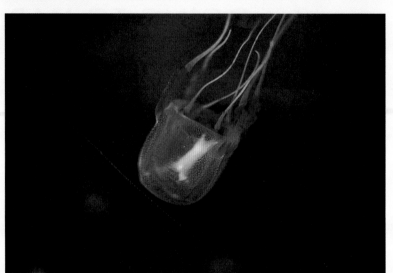

Photo: Dewald Kirsten/Shutterstock

Finally, they are directed to a cubicle where they find what they are seeking: underneath a mirror reflecting their own image is the label 'The most dangerous animal in the world'.

No one argues with the exhibitor's label: we all know that, collectively, we are responsible for more deaths than any other species on the planet. There is no doubt that we are capable of destroying vast areas of the land and sea along with much of Earth's biodiversity—but we are equally capable of looking after our terrestrial and marine environments, and nurturing their communities.

Not so long ago, the prevalent attitude towards the marine environment and its biodiversity was that they were so vast, and so varied, that any human effects on them would be negligible and of no significance. However, reports in the mass media reporting the devastating effects of habitat destruction and plastic pollution on our oceans have dispelled such notions for most of us.

There is an overwhelming consensus among the scientific community that threats to marine biodiversity from all kinds of human activities (highlighted in Figure 7.2) are greater today than ever before. This is linked to our ever-expanding global population demanding more coastal land and marine resources.

According to a UN SaveOurOcean document, in 2017 nearly 2.4 billion of the world's 7.5 billion people lived within 100 kilometres of the coast, putting unsustainable strain on coastal resources. The human population is projected to increase to more than 9 billion people by 2050, and this will only increase the threats to marine biodiversity.

Figure 7.2 The main human-generated threats to marine biodiversity.

Habitat destruction

The habitat is one of the defining characteristics of the ecological niche of a species. If a species has nowhere to live, it will fail to exist and become extinct. Habitat destruction is the most direct and obvious threat to marine biodiversity. It heads the IUCN hierarchical list of anthropogenic threats to biodiversity.

Marine habitats are destroyed directly by residential developments for housing and urban infrastructure; for tourist and recreation areas; and by commercial developments, including harbour installations for fishing and shipping. Marine habitats are also destroyed indirectly by human activities, such as fishing, often as an unintended consequence (see Case study 7.1).

Case study 7.1
Bottom fishing: from one extreme to the other

Fishing from large bottom trawlers and dredgers is not done with the intention of damaging habitats and destroying marine life. This is simply the inevitable consequence of fishing gear being dragged across the sea floor. We are going to look at two examples of such fishing: stern trawling for benthic fish and hand dredging for oysters.

Bottom trawling is a widespread commercial fishing technique which commonly involves dragging heavy nets, metal doors, and chains over the sea floor to catch fish (Figure A). This form of trawling disturbs the natural sea floor habitat, sometimes to the point of destruction, and can result in huge losses of marine biodiversity. However, the amount of damage caused varies with local conditions and the type of trawling. Some trawls skim just over the seabed

Figure A Bottom trawling from a fishing boat in which a net and metal plate is dragged along the sea floor behind a boat on the surface. The inset shows how this can result in an increase in turbidity when the metal door disturbs soft sediment.

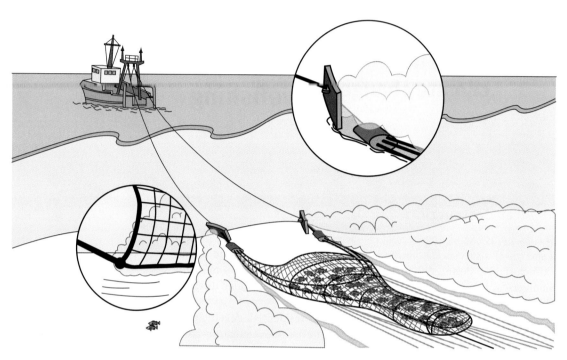

Credit: Ferdinand Oberle. Public domain

with minimal contact, others are designed to disturb animals that are buried in the first few centimetres of the seabed. Trawling also takes place over different substrates—the sea bed varies from hard rocks to fine sands and mud.

Bottom trawling can be particularly destructive when carried out in deep water because it requires a large fishing vessel with very heavy gear to catch benthic fish. Typically, deep-sea trawling gear includes an otter trawl that uses metal 'doors' weighing up to 5000 kilograms in order to get the net to the bottom and keep the net mouth open while being pulled across the sea floor. The trawl can be very wide, with the total width measuring up to 200 meters. The ground gear of a deep-sea trawl is equipped with rollers made of metal or stiff rubber that enable the net to move over rough substrates without getting tangled up. The heavy trawling gear is designed to maximize fishing efficiency in the deep sea. However, it is also potentially very destructive as it can disturb and even destroy benthic habitats and species. Corals and sponges are especially vulnerable to deep-sea trawling.

Despite its potential for destruction, supporters of bottom trawling argue that, if carried out properly with the correct gear and in a suitable habitat, it can be a sustainable form of fishing that has no long-term impact on the benthos.

In striking contrast to the type of fishing carried out by deep-sea trawlers is the hand-dredging for wild native oysters (*Ostrea edulis*) carried out in the beautiful waters of the Fal estuary in Cornwall (Figure B).

In 1876, local byelaws prohibited the mechanized harvesting of oysters in the Fal. These byelaws are still in force. Although oyster dredgers are able to use motors to get to the oyster beds, once in the beds they have to use sail to move about, and hand dredges to catch their quarry.

The dredges are made of steel and are up to 1.2 meters wide. A flat, stainless steel blade scrapes oysters and other benthic organisms from the seabed. Behind the blade, a light chain-link belly backed by netting and a hardwood batten retains the catch. Each dredge is attached to a rope let out over the side of the dredger, and hauled back in after the catch is made. Oysters are retained while other species caught in the net at the same time (the by-catch) are returned to the sea.

The movement of the sail-powered oyster dredgers over the fishing ground is governed by a combination of wind and tide.

As well as being limited to wind and muscle power, dredges must not exceed 20 kg in weight. Dredging is permitted only between 9am and 3pm on weekdays, between 9am and 1pm on Saturdays, and no fishing at all is permitted on Sundays. It is generally acknowledged that these restrictions have helped the Fal River oyster fishery become one of the most sustainable fisheries in the world.

Figure B Falmouth oyster dredging—Fal oystermen are the last in Europe to fish commercially under sail. They use gaff-rigged cutters, some over a century old.

Photo courtesy of Christopher Jones

❓ Pause for thought

There is a view that all bottom fishing using heavy gear should be banned. There is a contrary view that bottom trawled fish are an invaluable source of protein and nutrients for an ever-expanding world population (in 2018, bottom trawling was responsible for a quarter of all fish caught). Discuss the pros and cons of both views.

Which feature of Falmouth oyster dredging do you think is the most important in minimizing damage to the sea floor and marine biodiversity? What factors, other than habitat destruction, threaten the Falmouth oyster fishery?

Climate change and ocean warming

If habitat destruction is the greatest actual threat to marine biodiversity, climate change takes the top spot in most polls for being the greatest potential threat.

The 2019 IPCC *Special Report on the Ocean and Cryosphere in a Changing Climate* stated that climate change due to past and current emissions of greenhouse gases is causing profound changes in the marine environment (Figure 7.3): the global ocean is becoming warmer, more acidic, and less productive.

Climate change threatens marine biodiversity in many ways. One effect—probably the most alarming—is the warming of the polar seas, which in turn is causing glaciers and ice sheets to melt and sea levels to rise.

With continued climate change, extreme events, such as storms and high waves, are becoming more severe. The rising sea level and storm damage associated with climate change is threatening the very existence of low-lying coral atolls, such as those in the Maldives and Tuvalu. Atolls with healthy corals are able to protect themselves by accumulating sediment in response to the threats. But those weakened by bleaching events (see Section 7.3) haven't got the energy to do that.

The most direct impact of ocean warming on marine organisms is thermal stress. Species such as *Limacina helicina* (see Case study 7.2), living in the high latitudes of Antarctic waters, typically experience a narrow range of relatively low temperatures. They are generally more vulnerable to changes in sea temperature than species at middle or low latitudes, which normally experience much broader temperature ranges.

The rising temperature of a warming ocean increases the metabolic rate of ectothermic organisms, which leads to an increased demand for food. At high latitudes, ocean warming during winter may increase energy demands of organisms when primary production is low. This imposes a particular threat on species such *as Limacina helicina* which do not stop growing during the wintertime.

Ocean warming is not a new phenomenon. Cycles of cooling and warming have occurred since the oceans were first formed. Given time and a slowing down of ocean warming, many marine organisms can become

Figure 7.3 Important abiotic changes associated with anthropogenic climate change: human activities such as burning fossil fuel lead to an increase in concentration of greenhouse gases in the atmosphere; this in turn leads to other physical and chemical changes in the marine environment.

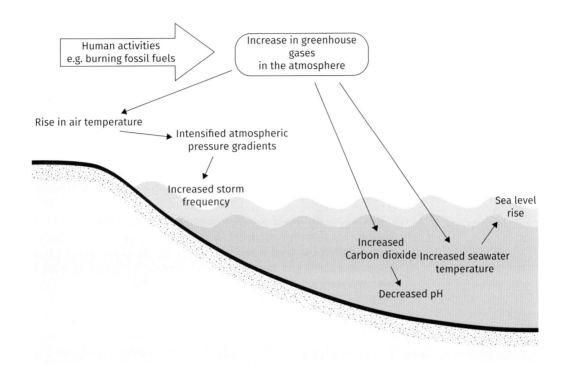

Source: Harley, C.D.G., Randall Hughes, A., Hultgren, K.M., Miner, B.G., Sorte, C.J.B., Thornber, C.S., Rodriguez, L.F., Tomanek, L. and Williams, S.L. (2006), The impacts of climate change in coastal marine systems. Ecology Letters, 9: 228–241. https://doi.org/10.1111/j.1461-0248.2005.00871.x

adapted to thermal stress. But what is different, and most concerning, about the current period of global warming is that it is occurring at such a fast rate.

Coral bleaching

Coral bleaching results from symbiotic algae (see Figure 1.4) evacuating coral tissue. Without algae, the coral loses its major source of food, turns white or very pale, and becomes more susceptible to disease and other environmental stressors. Although corals can survive a single bleaching event, repeated bleaching increases the risk of death.

The leading cause of coral bleaching is an increase in sea temperature caused by anthropogenic climatic change. Other causes are freshwater runoff and pollution, overexposure to sunlight, and extreme low tides; the effect of all of these is amplified by thermal stress.

According to a 2019 IPCC Special Report on climate change, '*Marine heat waves have doubled in frequency since 1982 and are increasing in intensity. They are projected to further increase in frequency, duration, extent, and intensity. Their frequency will be 20 times higher at 2 °C warming, compared to pre-industrial levels. They would occur 50 times more often if emissions continue to increase strongly.'*

Ocean deoxygenation and dead zones

Climate change in combination with nutrient pollution is responsible for ocean deoxygenation, a loss of oxygen dissolved in the seawater of the marine environment. It threatens all aerobic organisms, especially fast-swimming pelagic predators, such as squid, tuna (Figure 7.4), marlin, and sharks, which all need well-oxygenated waters to survive.

Ocean warming linked to anthropogenic climate change leads to ocean deoxygenation because the solubility of oxygen in seawater decreases with temperature rise (see Chapter 3). It is estimated that between 1960 and 2010 oxygen levels in the ocean declined globally by about 2 per cent, with the decline in some tropical regions being as high as 40 per cent. Much of the oxygen loss occurs in the top 1000 metres of tropical seas.

Minna Epps (IUCN's Global Marine and Polar Programme Director), being mindful that slow-moving jellyfish thrive better than active predators in oxygen-depleted waters, is reported as saying, '*If we run out of oxygen it will mean habitat loss and biodiversity loss and a slippery slope down to slime and more jellyfish.'*

When oxygen depletion reaches a critical level, it results in a hypoxic area or dead zone which does not have sufficient oxygen to support aerobic marine life. Unfortunately for us and our attempts to control climate change, anaerobic bacteria thrive in hypoxic conditions; some species

Figure 7.4 Tuna, one of the species particularly threatened by ocean deoxygenation.

Photo: James Thornton/Unsplash

release large volumes of nitrous oxide, a gas with a greenhouse effect 300 times that of carbon dioxide. Dead zones are linked to climate change, but they can be caused by any oxygen-depleting factor. Historically, human pollution has been a major cause of dead zones, especially in coastal areas.

The Gulf of Mexico is a major coastal area regularly depleted of oxygen. Each year, a combination of high spring rainfall, river discharge (mainly from the Mississippi watershed), and nutrient pollution causes a dead zone covering around 14 000 square kilometres. Although vast, this dead zone is only the second largest in the world. The largest occurs in the Arabian Sea. Its area is more than 160 000 square kilometres, covering almost the entire Gulf of Oman.

Ocean acidification

An increase in atmospheric carbon dioxide results in more of the gas being dissolved in seawater and a lowering of oceanic pH. The ocean has absorbed between 20 and 30 per cent of human-induced carbon dioxide emissions since the 1980s, causing widespread ocean acidification.

Ocean acidification tends to reduce carbonate concentrations in seawater and inhibit calcification (the formation of calcium carbonate). As a result, it becomes more difficult for many marine organisms to incorporate calcium carbonate into their skeletons. In addition, if the pH becomes too low, calcium carbonate dissolves slowly as a result of a decrease in saturation levels. This is more pronounced in the cold seawaters of polar regions and the marine environments off the west coasts of continents, where saturation levels are lower than in warm waters.

Ocean acidification is a particularly important threat to coral reef ecosystems. Their extraordinary biodiversity relies on the persistence of carbonate structures in coral skeletons. Ocean acidification reduces the ability of coral to use calcium carbonate to build new skeletal material, and it increases the rate at which skeletal structures already formed dissolve away. This results in coral reefs going from a condition of net building to one of net erosion.

Coccolithophores, crustacea, and molluscs (especially pteropods – see Case study 7.2) are among other marine calcifiers prone to shell dissolution.

Cumulative effects of stress

The stresses due to ocean warming, acidification, and deoxygenation are wide-ranging and often occur together in the marine environment (Figure 7.5). Any marine organism has to be able to cope not only with individual stresses but also with stresses acting together. Most of us know only too well that, however admirably we might cope with a single stress, when stresses pile up the addition of just one more may be 'the straw that breaks the camel's back'.

Changes in ocean circulation

Some of the most important questions in marine science concern how climate change affects ocean currents and the great Ocean Conveyor Belt (see Figure 2.14). Probably one of the most studied movements of oceanic waters is the Atlantic Meridional Overturning Circulation (AMOC; Figure 7.6).

Case study 7.2
Sea butterfly under threat by climate change

Shelled pteropods such as *Limacina helicina* (see Figure A), also known as sea butterflies, secrete mucus webs up to 5 centimetres in diameter, several times larger than themselves, to trap phytoplankton. The web also acts as a buoyancy aid to keep the animals afloat when stationary. With no mucus web, and when not flapping their 'wings', sea butterflies sink to the sea floor. They are entirely pelagic.

Pteropods spend their entire lives as plankton in the world's oceans. They are regarded as a keystone species complex, especially in Arctic waters where they occur in enormous numbers per unit volume and can comprise more than 50 per cent of total zooplankton abundance.

Limacina helicina and related marine molluscs play an important role in carbon sequestration by consuming huge quantities of phytoplankton, and by being consumed by fish. When sea butterflies die, or when their faecal pellets and mucous food strings sink, they transfer carbon to the ocean floor. A decline in population abundance of sea butterflies could lead to a dramatic increase in ocean acidification. This is an especially important threat in the Southern Ocean, because this body of water is currently responsible for about 50 per cent of the total global carbon dioxide uptake.

Sea butterflies are affected by a triple whammy of climate change impacts: ocean warming, acidification, and oxygen depletion (see Figure 7.5).

The shells of pteropods are made mainly of aragonite, a stable form of calcium carbonate; they are light and transparent to allow for a pelagic life

Figure A *Limacina helicina* is a minuscule marine mollusc, less than 0.5 millimeters long. It belongs to the shelled pteropods, commonly known as sea butterflies because of their graceful, winged movements.

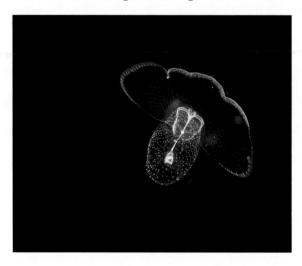

Photo: Peter Leahy/Shutterstock

and camouflage. Ocean acidification, a lowering of seawater pH, increases the rate at which the shell dissolves and inhibits calcification. If dissolution exceeds calcification, shells become thinner, more fragile, and more porous, making them more vulnerable to predation and other deadly threats.

Pteropods typically change their depth distribution at different times of the day, a phenomenon called **Diel Vertical Migration** (DVM). They tend to rise towards the surface layer at night and descend to deeper waters during the day. There is evidence that some shelled pteropods respond to ocean acidification by increasing their movements up and down the water column so that they can minimize time spent in areas where carbon dioxide levels are highest.

? Pause for thought

What are the costs and benefits of sea butterflies moving towards the surface at night and down into the deeper levels during the day?

Figure 7.5 The impact of multiple stresses resulting from climate change on shelled pteropods (sea butterflies). Many, if not most, marine organisms face similar multiple stresses in their environment.

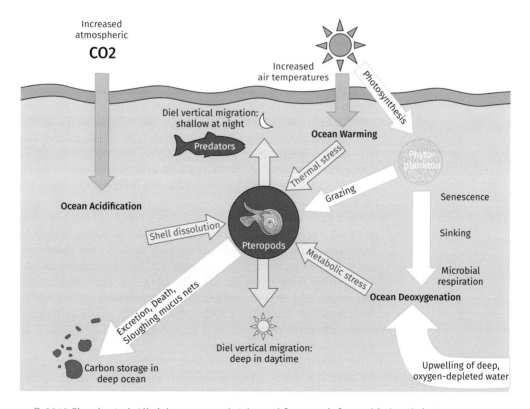

Figure 7.6 The Atlantic Meridional Overturning Circulation (AMOC) transports heat northwards, contributing to the United Kingdom's mild climate. Red colours indicate warm, shallow currents and blue colours indicate cold, deep return flows.

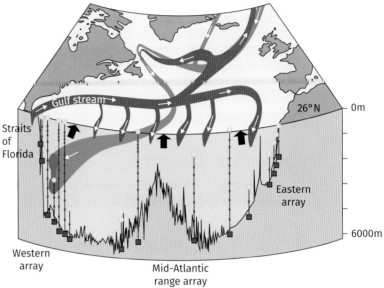

Source: © Copyright CSIRO Australia/Louise Bell and Neil White

The AMOC is driven mainly by the Gulf Stream that flows from the Straits of Florida to temperate waters, transferring heat as it goes. It is such an important carrier of thermal energy that the Marine Climate Change Impact Panel (MCCIP) regard it as the key to understanding any trends and effects of global climate change on the UK climate and its marine biodiversity. A 2020 MCCIP report stated that changes in the strength of the AMOC have led to several decades of changes in the climate in the UK, causing summers to become generally drier. The report predicts that the AMOC will weaken during the remainder of the twenty-first century due to climate change, potentially causing large biogeographical and climatic shifts that will affect UK marine biodiversity.

These shifts are already happening. MBA scientists Dr Nova Mieszkowska and Professor Stephen Hawkins, and their team of marine biologists have demonstrated that Lusitanian species, such as the limpet *Patella depressa* and the brown seaweed *Bifurcaria bifurcata*, both originating in the warm waters to the south of the UK, are expanding their distribution into higher latitudes around the UK coastline. In contrast, boreal species, such as the acorn barnacle *Semibalanus balanoides* and kelp *Alaria esculenta* which originate from the cooler waters of the Arctic, are shifting their distributions northwards, away from the UK coasts.

The 2004 film *The Day after Tomorrow* features a sudden collapse of the AMOC caused by global warming. It leads to freezing conditions with

New York being completely iced over. Such a freezing would have dramatic effects on marine biodiversity. In their 2020 Report Card (see Scientific approach 7.1), the MCCIP concluded that, in general, climate models do not predict an abrupt shutdown of the AMOC this century. The disasters depicted in *The Day after Tomorrow* are not imminent—but the slowing of the AMOC is still a cause for concern.

Scientific approach 7.1
Climate change: examining the evidence

The Marine Climate Change and Impacts Partnership (MCCIP) was created in 2005 to collect, collate, synthesize, and report on evidence about marine climate change impacts for the UK. Its reports, such as the one on sand eels shown in Figure A, are used to inform policy makers and management on the impact of climate change on the UK marine environment and its biodiversity.

The MCCIP is a collaboration between marine scientists and sponsors from the UK government agencies, NGOs, and industry. Report cards, such as that on sand eels, are based on peer-reviewed papers from many experts, and information contained in previous reports.

To ensure scientific integrity and to demonstrate independence, the authors of the MCCIP reports adopt a robust and transparent four-step process that mitigates against accusations of bias in reporting (Table 1).

The Report Cards are used to inform policy makers and management on the impact of climate change on the UK marine environment and its biodiversity. For example, the 2018 MCCIP Report Card gave the following key points about sand eels and their availability as seabird prey in the context of climate change:

- Sandeels are an important trophic link between plankton and predatory fish, seabirds, and mammals, and support a large fishery in the North Sea.

- Seabirds are particularly sensitive to sandeel availability because they depend on them to feed their chicks.

- Climate change can have a direct impact on the reproductive timing of sandeels and the **phenology** of the plankton prey they depend on, increasing the likelihood of a mismatch between sandeel larvae and their prey, leading to poor recruitment.

- Current approaches to managing sandeels include population level landing restrictions and closed areas. However, further restrictions on anthropogenic activities in seabird foraging areas could be considered. The growing contribution of alternative prey such as sprat requires that fisheries on forage species should take account of predator requirements.

Figure A The cover of a 2018 MCCIP report on climate change and marine conservation. It focuses on sand eels (the fish held expertly in the beak of a puffin, just one of the predators that depend on sand eels for food).

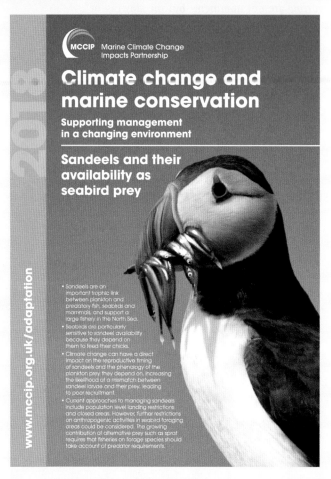

Image: © Crown Copyright 2021

Table 1 The processes used to ensure scientific integrity and independence of information given in MCCIP Report Cards

Step 1 Information identification		
Risk	**Mitigation**	**MCCIP approach**
Selection bias: 'cherry picking' topics or research areas that support pre-held opinions	• joint setting of 'information agenda' • transparent decisions	• The MCCIP Steering group comprising 26 partners identify the information needed

Continued

Table 1 *Continued*

Step 2 Expert Identification		
Risk	**Mitigation**	**MCCIP approach**
Expert bias: selecting a narrow group of experts known for promoting certain views or hypotheses	• Comprehensive expert involvement • Clear instruction to authors to include representative range of opinion • Independent peer review process	• Provisional lead authors identified and approached • Lead authors are required to represent and work with community of experts in their field regardless of individuals' opinions • Materials produced by authors are anonymously and independently peer-reviewed and revised accordingly

Step 3 Information translation		
Risk	**Mitigation**	**MCCIP approach**
Interpretation bias: those responsible for translating the information can introduce their own bias and opinion	• Clear terms of reference and accountability • Scientists cross check • Information and data audit	• Report Card Working Group established – individuals, not mandated as experts, not representatives of their organizations • All summary information and data made publicly available (online) and any publications provided as open access in journals

Step 4 Information communication		
Risk	**Mitigation**	**MCCIP approach**
Evidence 'weighting' bias evidence or advice may be given too much credence or credibility	• Confidence assessment	• Lead authors provide confidence rating as indication of uncertainty around topic • Simple language used to avoid ambiguity

Modified from Frost et al. (2018) Reporting marine climate change impacts: lessons from the science—policy interface; in MCCIP (see Further reading).

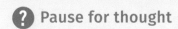 **Pause for thought**

When giving information in scientific reports, which risk to scientific integrity and independence is the greatest? Which risk is least?

Marine pollution

Marine pollution is defined by the United Nations Joint Group of Experts on the Scientific Aspects of Marine Pollution (GESAMP) as the '*Introduction by man, directly or indirectly, of substances or energy into the marine environment (including estuaries) resulting in such deleterious effects as harm to living resources, hazards to human health, hindrance to marine activities including fishing, impairment of quality for use of sea water and reduction of amenities*'. As such, marine pollution is a major threat to marine biodiversity. However, the global ocean is so vast that it's difficult to get out of the mindset that we can dump almost anything into it with impunity: seawater will just dissolve all of our troublesome waste away. Much of the world's waste, estimated to be about 20 billion tons per year, ends up in the sea, often without any preliminary processing.

There are many types of pollutants. They include oil, plastics, and litter (Figure 7.7); toxic chemicals such as tributyl tin, acids, and pesticides; organic effluents that cause eutrophication which can lead to algal blooms; and energy pollution in the form of light, noise, radioactive substances, and thermal pollution. Each type of pollution threatens marine biodiversity in its own particular way. For example, polychlorinated biphenyls (PCBs) are a group of man-made, very stable chemicals used widely in electrical equipment such as capacitors and transformers. They may appear to be of little consequence when they first enter the ocean, but biomagnification results in their concentration increasing up the food web so that they become a very real threat to top predators, as Figure 7.8 clearly shows.

Figure 7.7 Fishermen working amidst floating garbage, mainly plastic, in Manila Bay in the Philippines. It is close to the Great Pacific Garbage Patch, which has the largest accumulation of surface litter in the global ocean.

Photo: Reuters / Alamy Stock Photo

Figure 7.8 Biomagnification of polychlorinated biphenyls (PCBs) up the food web.

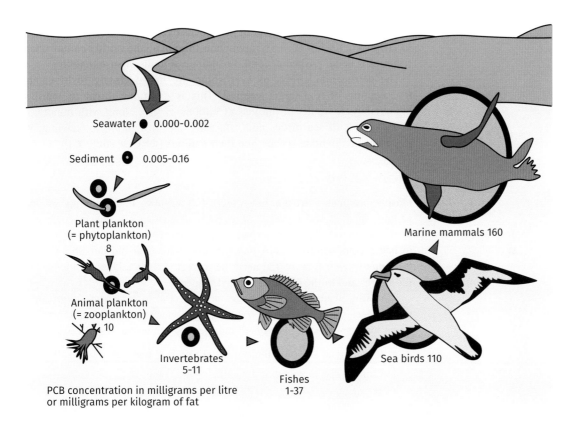

Source: Walter Maria Scheid, Berlin, for World Ocean Review 2010

Probably the marine pollutant with the worse publicity in recent years is plastic. The use of plastic has been criticized to the point of demonization in some of the mass media. There is no doubt that its persistence, its ability to block the airways and digestive system of marine organisms, and its role as a reaction surface for the production of chemicals dangerous to marine organisms and humans, make plastic a very serious threat. However, it should not be forgotten that plastic also has a very positive role to play in our lives, and in marine conservation.

The main problem with plastic products is not that we use them per se, but that we use them indiscriminately, and we don't discard them with due care and attention. Most campaigners motivated by the appalling amount of plastic that enters the marine environment, target single-use plastics, in cups, lids, straws, and plastic bags, as things that we should and can do without.

Overfishing

In the context of the marine environment, overfishing refers to harvesting 'fish' from a body of water at rates too high for the fished species to replace themselves. 'Fish', in the context of overfishing, generally refers to elasmobranch (cartilaginous) and teleost (bony) fish, crustaceans, molluscs, and other aquatic animals except mammals and reptiles.

According to the FAO report *The state of the World Fisheries and Aquaculture 2020*, the number of fish stocks overfished in the global ocean has tripled in the last 50 years; about one-third of all the world's commercial fisheries is being pushed beyond the biological limits of sustainability.

The detrimental effects of overfishing are linked closely to by-catch (the capture of non-target species). This is a serious threat to marine biodiversity. It results in the loss of billions of fish, along with hundreds of thousands of sea turtles and cetaceans and, in relation to trawling, it threatens invertebrate species and their habitats (see Case study 7.1).

Non-native species

Marine organisms are no respecters of international boundaries set by people. Japanese wire weed, *Sargassum muticum*, is a species endemic to the waters of Japan. It needed no passport to enter British waters and to begin colonizing our rock pools in the 1970s. Nor did our Shore crab, *Carcinus maenas* (Figure 3.3) need the correct visas to take up residence on the shores and in the estuaries of the USA.

The MBA Bishop Group provides a guide and information about non-native species in the UK (at www.nonnativespecies.org). The Group identifies seven main sources and vectors of marine non-native species (NNSs):

- Commercial shipping (ballast water and hull fouling)
- Commercial movement of shellfish
- Movement of service barges, towed hulks, and pontoons
- Movement of leisure craft
- Release of aquarium species
- Shipment of angling bait
- Importation of research material.

Many marine NNSs have become a threat to so-called native indigenous marine organisms, and a pest to human activities. For example, in UK waters, *Sargassum muticum* can grow so prolifically it outcompetes some native species. In the USA, *Carcinus maenas* is a predatory threat to indigenous molluscs and a pest of shellfisheries. Other NNSs that pose a threat to endemic species include sea-squirts, sea-mats, barnacles, and bivalves (Figure 7.9).

Protecting marine biodiversity

As well as having the power to destroy marine biodiversity, we humans also have the power to protect it. There are numerous conservation organizations, from the local to the global. Every member state of the United Nations has government representatives responsible for implementing the Convention on Biological Diversity (discussed at the beginning of Chapter 4). In Europe, the Habitats Directive (also mentioned in Chapter 4) has been instrumental in providing a framework for conserving marine biodiversity.

There is a variety of government and nongovernment bodies that are committed to promoting marine conservation policies and legislation

Figure 7.9 Just some of the non-native species of barnacles, molluscs, sea squirts, sea mats, a sea anemone, and a polychaete worm that have been introduced onto UK shores and threaten marine biodiversity. How many can you identify?

Image: John Bishop © Marine Biological Association

designed to sustain our fisheries and other living resources. Every policy and every piece of legislation, such as fish quotas and no-take zones, has its costs and benefits (see Figure 7.10). These need to be evaluated carefully before policies are implemented, and reviewed during implementation

Figure 7.10 The benefits of adopting a no-take zone.

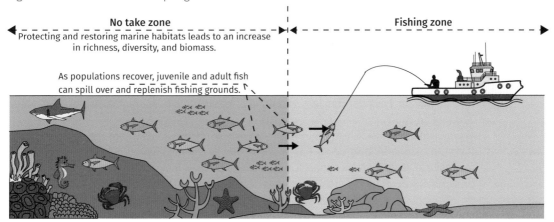

Source: Aning Kuswinarni

to ensure they are effective, achieve the stated aims, and do not deliver unintended and damaging consequences. No-take zones are important components of many Marine Protected Areas (MPAs) (see The bigger picture, 7.1).

Advocating marine protection

Many different people around the world are raising awareness of the urgent need to protect and treasure marine biodiversity.

Sir David Attenborough (Figure 7.11) is probably the most famous advocate for marine protection.

Sir David's commentary for the *Blue Planet* series of documentaries has had an enormous impact on the public understanding of human threats to the global marine environment and biodiversity.

Other advocates for marine protection include young marine biologists, such as Amanda Hodo (Figure 7.12), a biologist at the Mote Marine Laboratory and Aquarium in Sarasota, Florida.

Figure 7.11 Sir David Attenborough.

The bigger picture 7.1
Marine networks: more than the sum of their parts?

There are about 5000 MPAs worldwide. The largest is in the Ross Sea region of Antarctica. It was created in 2017 and covers an area of more than 1 500 000 square kilometres.

Marine Protected Areas (MPAs) are generally accepted as an essential tool for reversing harmful anthropogenic effects on marine biodiversity. In a paper published in the *Marine Ecology Progress Series* in 2009, Sarah Lester and her co-authors reviewed 124 'no-take' MPAs in 29 countries and reported that on average they had 21 per cent more biodiversity than adjacent fished areas. But it is important to know which types of MPAs are most effective.

Advocates of large MPAs say they benefit marine conservation more than small MPAs because their large size means that they can take advantage of economies of scale. This generally makes them more cost-effective than small MPAs. A single large MPA also allows different marine ecosystems to interconnect in ways not possible in a number of small and separate MPAs covering an equivalent area.

Critics of large MPAs claim that they are mainly in remote areas where the threat to marine biodiversity is low and that they are being created mainly for political reasons rather than for conservation. They maintain that the publicity surrounding the creation of massive MPAs distracts public concern away from the urgent task of protecting marine species in smaller areas that have actual and current threats.

The general scientific consensus is that MPAs of all sizes, if well managed and monitored, benefit marine biodiversity. Nevertheless, the optimal size of an MPA depends on which ecosystems and species are to be protected. For example, the optimal size of an MPA for highly mobile species, such as Giant squid, is generally larger than for less mobile species.

Enric Sala and Sylvaine Giakoumi carried out a meta-analysis of the effectiveness of MPAs (see Further reading). They reported that no-take marine reserves are more effective than MPAs that offer only partial protection: the biomass of all the fish species present in no-take marine reserves was, on average, 670 per cent greater than in adjacent unprotected areas, and 343 per cent greater than in partially protected MPAs. Their analyses also indicated that, once top predators have been re-established, fully protected marine reserves also help restore the complexity of damaged ecosystems through a chain of trophic cascade-like ecological effects. Although most marine reserves are conceived mainly to protect particular ecosystems within their boundaries, Sala and Giakoumi's meta-analyses also showed that they can enhance local fisheries and create jobs and new incomes through ecotourism.

Figure A An Australian giant cuttlefish (*Sepia apama*) in Spencer Gulf, part of the network of marine protected areas in South Australia.

Photo: The Commonwealth of Australia

The Giant cuttlefish (see Figure A) is protected in South Australia in a network of marine parks that include breeding and nursery areas which have been designated sanctuary zones (equivalent to fully protected MPAs). Collectively, the network of marine parks covers more than 40 per cent of the seas that come under the auspices of the South Australian Government. However, the combined area of the sanctuary zones amounts to only about 6 per cent of the area, and these sanctuary zones are in separate parts of the network

❓ Pause for thought

Suggest why the whole of the South Australia Marine Park Network has not been designated a sanctuary zone. Why are separate small sanctuary zones likely to be more effective at sustaining Giant cuttlefish than one large sanctuary zone taking up the same area (6 per cent) of the Marine Park Network?

In 2018, Amanda Hodo was named a 'hero' for conservation, education, and the environment by the Association of Zoos and Aquariums (AZA) for her commitment to marine animal care and conservation. She has become well-known in social media for her participation in a fish captive breeding programme that is part of a seahorse Species Survival Plan (see YouTube https://youtu.be/5V3ZJSFejnE).

Dr Sylvia Earle (Figure 7.13) is a distinguished underwater explorer, former chief scientist of the National Oceanic and Atmospheric Administration,

Figure 7.12 Amanda Hodo.

Figure 7.13 Dr Sylvia Earle talking to US President Barack Obama about marine conservation.

Photo: White House Photo / Alamy Stock Photo

and a tireless advocate for the protection of the marine environment and its biodiversity.

Sylvia has spent longer than 7000 hours underwater and led more than 100 scientific expeditions. So much of her life has been underwater, it has been said that she is probably more at home in the sea than on dry land.

Among her many commitments to marine conservation is her leadership of a nonprofit initiative called *Mission Blue*. Its aim is to 'ignite public support for a network of marine protected areas', such as those in South Australia (see The bigger picture 7.1). Sylvia Earle is an Honorary Fellow of the MBA (Marine Biological Association UK). In the 2014 Royal Charter edition of *The Marine Biologist* (an MBA publication), editor Guy Baker referred to her as:

> ... arguably the world's leading ocean advocate, in which the case for exploration, sharing knowledge, and improving the public perception of 'planet Earth's blue heart' is passionately and eloquently made.

In the Royal Charter edition of *The Marine Biologist*, Sylvia Earle and Dan Laffoley co-authored an article entitled 'Big, Blue and Beautiful', which explored why the Earth's 'blue heart' is the key to 'future wealth, health, and happiness'. Dan Laffoley is Principal Advisor for Marine Science at the IUCN, and another great advocate for the conservation of the marine environment and its biodiversity. The article finishes with the following piece which contains a message that I would like to share with you as we approach the end of *The Marine Environment and Biodiversity*:

> Now we understand that life exists from the surface to the greatest depths. And now we know that our lives depend on the ocean, and the creatures that live there. It is our life support system. It is a shocking truth but for every pound that goes to land conservation, birds, freshwater, whatever it is, the terrestrial parts of the planet, a penny goes for ocean protection. And that is a reflection of the attitudes people have. They understand that the forests and the rivers and the lakes are in trouble—and they are—so we should not take a penny away from protecting the land. We need much more devotion to both land and sea. We need to realise this is not a luxury. This is not an option. This is critical to our health and to our survival. We need to take seriously the importance of caring for the natural systems that take care of us.

Chapter summary

- The greatest current human threat to marine biodiversity is habitat destruction from activities such as fishing, housing, and industrial development. Habitat destruction is increasing with increases in global human population.
- Probably the greatest potential threat to marine biodiversity is from anthropogenic climate change. This is linked to ocean warming, ocean deoxygenation, and ocean acidification. Ocean warming can lead to coral bleaching and ocean deoxygenation.
- Climate change is particularly important in slowing down the Atlantic Meridional Overturning Circulation (the AMOC). The AMOC is not in imminent danger of stopping, but its slowing down is likely to have significant effects on regional climates and marine biodiversity.
- Marine pollution, overfishing, and the introduction of non-native species by human agents are all important threats to marine biodiversity.
- There are many examples of human actions and organizations that help protect marine biodiversity, but well-designed and well-managed Marine Protection Areas are probably the most effective.

Further reading

Henson, R. (2019) *The thinking person's guide to climate change.* 2nd edn. American Meteorological Society.
A readable, authoritative account of climate change. An excellent introduction to a very controversial subject. It covers the effects of climate change on marine environments and biodiversity.

Hiscock, K. (2014) *Marine biodiversity conservation: A practical approach.* Routledge.

A useful resource for anyone aspiring to be a marine conservation biologist.

MCCIP (2018) *Marine climate change impacts: 10 years' experience of science to policy reporting.* (Eds Frost, M., Baxter, J.M., Buckley, P., Dye, S., and Stoker, B.) Summary Report MCCIP Lowestoft 12 pp doi:10.14465/2017.arc10.000-arc.

A report that summarizes UK research on the effects of climate change on marine biodiversity up to 2017.

Sala, E., and Giakoumi, S. (2017) No-take marine reserves are the most effective protected areas in the ocean. *ICES Journal of Marine Science,* 75, 1166–1168.

A succinct and interesting evaluation of the effectiveness MPAs; available as a pdf.

Weis, J. S. (2014) *Marine Pollution: What Everyone Needs to Know.* Oxford University Press.

An interesting introduction to key issues about marine pollution.

 Discussion questions

7.1 What are the advantages and disadvantages of having materials made of biodegradable plastics and very long-lasting plastics?

7.2 Alternative terms for non-native species are 'alien species' and 'invasive species'. Suggest why some marine conservationists dislike the use of the alternatives. Which do you think is the most appropriate term?

7.3 Is there any difference between conservation, preservation, and protection of marine environments and their biodiversity?

GLOSSARY

abiotic conditions Due to physical, non-biological environmental factors such as temperature, salinity, and pH.

abyssal benthic zone The deepest part of the ocean between 3000 meters and 6000 meters below sea level.

abyssal plains Relatively flat and muddy areas that make up about two thirds of the sea floor.

abyssobenthic zone Alternative term for abyssal benthic zone.

accessory photosynthetic pigments Any of various light-absorbing compounds that transfer energy from that light to the primary photosynthetic pigment for photosynthesis.

aerobic chemosynthesis Process using oxygen in seawater to oxidize inorganic compounds and release energy to make organic compounds from carbon dioxide and water.

algae Informal term for a large and diverse group of eukaryotic, photosynthetic organisms ranging in size from single-celled forms to large seaweeds.

algal bloom Rapid increase in the population of algae.

alpha diversity Biodiversity within a particular area or ecosystem, expressed by the number of species occurring there.

amphidromic point Location in the sea around which a tidal system oscillates, and where there is no change in the height of the water.

anoxygenic photosynthesis Type of photosynthesis in which no free oxygen is released; water is not split; a different molecule such as hydrogen sulfide is used as an electron donor.

aphotic zone Region of the marine environment where no sunlight penetrates.

applied research A systematic study undertaken mainly to solve a practical problem, for example fisheries.

archaea Formerly known as archaebacteria, single-celled organisms with distinct molecular characteristics that separate them from true bacteria.

assemblage A collection of organisms characteristically associated with a particular marine environment.

Atlantic Meridional Overturning Circulation A large system of ocean currents characterized by a northward flow of warm, salty water in the upper layers of the Atlantic, and a southward flow of colder water in the deep Atlantic.

autopoiesis Refers to a system able to reproduce and maintain itself.

bacteriophage Type of virus that infects and destroys bacteria.

bathyal zone Part of the open ocean that extends from a depth of 1000 metres to 4000 metres below the sea surface.

benthic zone In the marine environment, the lowest level consisting of the sediment surface, the water immediately above it, and some subsurface layers.

beta diversity Refers to a comparison of the biological diversity between two distinct ecosystems or geographical areas.

bioaccumulation A progressive increase in the concentration of substances, such as pesticides, in organisms that live in polluted environments.

biological carbon pump A collection of biological and physical mechanisms that transport carbon from the ocean surface to its depths. It refers mainly to the fixation of carbon dioxide into particulate organic carbon and its subsequent transfer to the deep ocean through gravitational sinking.

biological clock An internal 'molecular time piece' controlling periodic processes such as reproductive behaviour.

Biological Species Concept The view that a species comprises a population (or groups of populations) whose members are only able to interbreed with each other to produce fertile offspring.

bioluminescence The emission of visible light by living organisms.

biome A major ecological community or group of communities, generally characterized by a dominant type of vegetation, that extends over a large geographical area.

blowhole In whales and dolphins, the nostril on the top of the head.

boreal species Organisms located in northern waters; in marine biology, typically cold-water species.

brine A concentrated solution of salts which can be formed by the partial evaporation of seawater.

calcifiers Organisms that incorporate calcium carbonate into their shells or skeletons.

carbon sequestration The removal or capture of carbon dioxide from the atmosphere.

climate change mitigation Refers to actions that reduce or prevent the emission of greenhouse gases.

cohesion The sticking together of particles that are the same, such as water molecules.

compensation point The light intensity at which respiration and photosynthesis are in balance.

continental shelf Gently sloping edge of a continent that lies under the ocean.

continental slope A steep slope between the outer edge of the continental shelf and the deep ocean.

convergent evolution The development of superficially similar structures in unrelated organisms, usually because the organisms live in the same kind of environment.

coral bleaching The whitening of coral which occurs when coral expels zooxanthellae (symbiotic algal cells); it usually indicates environmental stress.

Coriolis effect The apparent tendency for objects, including oceanic and atmospheric currents, to be deflected to the right in the northern hemisphere and to the left in the southern hemisphere.

cultural services Non-material benefits that people obtain from ecosystems.

degassing The release of gases from the Earth's interior during volcanic activity.

dendrogram A type of tree diagram that shows the hierarchical relationship between items, such as genes and species.

density A measure of mass per unit volume.

deuterostome A subdivision of the animal kingdom that includes echinoderms and chordates; they share a common embryonic development in which the opening in the embryo develops into the anus; the mouth forms as a secondary opening.

Diel Vertical Migration In marine organisms such as zooplankton and fish, a synchronized pattern of movement up and down in the water column over a daily cycle.

dipolarity Applied to a molecule, such as water, that has a positively and a negatively charged end.

dissolved organic matter Material in seawater that contains organic carbon from a variety of biological and geological sources; generally taken as organic matter that passes through a filter of pore size 0.22–0.7 micrometres.

DNA barcoding A method of identifying an organism using a short section of DNA from a specific gene or genes.

DNA primer A short single-stranded nucleic acid that initiates DNA synthesis.

drag Force opposing the motion of a body through a gas or liquid.

dynamic soaring A flying technique used by birds to gain energy by repeatedly crossing the boundary between air masses moving at different velocities.

dysphotic zone The region in the marine environment where light intensity is such that oxygen production by photosynthesis is exceeded by oxygen consumption by respiration.

ebb tide Period between high and low tide, when the sea level falls.

ecosystem diversity The variation in the ecosystems found in a region.

ecosystem sentinel A species that responds to changes in an ecosystem in a timely, measurable, and interpretable way.

ectotherm An animal that has a body temperature dependent on the heating and cooling effects of the external environment.

eDNA See environmental DNA.

Ekman spiral A theoretical model that explains how the direction of ocean currents is displaced relative to the direction of a steady wind blowing over the sea.

Ekman transport The tendency of masses of oceanic water to move in a direction at right angles to the prevailing wind.

elasmobranchs Cartilaginous fish that comprises sharks, skates, and rays.

elucidating sentinels Organisms that act as indicators of past or ongoing ecosystem changes that would otherwise go unnoticed.

environmental DNA eDNA, genetic material containing DNA obtained directly from environmental samples (e.g. seawater and marine sediments) without any obvious signs of the source of the material.

epifluorescence microscopy A technique that uses a high intensity light source (such as a mercury lamp) to illuminate a sample from above a specimen and an optical system that enables only light emitted by fluorescent molecules in the specimen to be observed or photographed.

euphotic zone The uppermost layer of a body of water where there's enough light for the rate of photosynthesis to exceed the rate of respiration in photosynthetic organisms.

fetch The distance over which the wind acts upon the sea surface.

flood tide Period between low tide and high tide when the sea level rises.

functional biodiversity Type of biological diversity based on the range of things organisms do within an ecosystem or geographical area.

functional trait An observable characteristic or feature of an organism that determines the effect of the organism on ecological processes and its response to environmental factors.

gamma diversity The overall species diversity of a range of ecosystems, habitats, or communities within a region.

gel electrophoresis A technique used to separate macromolecule fragments (especially those of DNA, RNA, and proteins) according to their size.

genetic diversity A measure of biodiversity based on the number of different types of genes in a species, population, or the community of an ecosystem.

Green Fluorescent Protein GFP, a protein isolated from certain jellyfish and sea anemones that emits a green fluorescence when exposed to light in the blue to ultraviolet range.

gyre Any large system of rotating ocean currents.

habitat Place in which an organism normally lives, characterized by its physical and biological conditions.

hadal zone The deepest part of the marine environment, 6000 metres below the sea surface.

heterozygosity index A measure of the number of gene loci that are heterozygous; a completely inbred population, homozygous at all loci, would have a heterozygosity index of zero.

High Nutrient–Low Chlorophyll region HNLC region, a region in which a spring algal bloom is rarely observed and the level of macronutrient (nitrate) is rarely depleted.

histochemical staining A process used to study the chemical constituents of a biological material (e.g. a cell or tissue) by means of staining agents.

homeostasis The tendency of a biological system to resist change and maintain a relatively steady and stable state, especially by means of physiological processes.

hotspot Applied to biodiversity, a region that has an exceptionally high diversity of organisms threatened by human activity.

hydrogen bonding A relatively weak force of attraction between two molecules resulting from the electrostatic interaction between a proton (H^+) in one molecule and an electronegative atom (e.g., oxygen) in the other molecule.

hydroskeleton A type of exoskeleton supported by the pressure of a watery fluid.

ionic Applied to a chemical bond that is formed by the electrostatic force of attraction between a positively charged metal ion and a negatively charged non-metal ion.

intertidal zone Region of the marine environment that is periodically exposed to the air at low tide.

irradiance Light energy falling on unit area of a horizontal surface in unit time.

keystone species A species that has a disproportionately strong effect on a particular ecosystem, such that its removal results in severe destabilization of the ecosystem and can lead to further species losses.

latent heat of fusion The amount of heat energy required to change a substance from solid to liquid at constant temperature (also known as latent heat of melting).

leading sentinel An organism used to predict future changes in an ecosystem; typically, a leading sentinel has a lower threshold to changes in the environment.

Liquid Chromatography-Mass Spectrometry LC-MS, an analytical technique that combines the physical separation capabilities of liquid chromatography with the power of mass spectrometry to detect, identify, and quantify molecular components in a sample.

Lusitanian species In marine biology, applied to species originating from the warm waters of Portugal and Spain.

marine environment The global ocean or part of it.

Marine Protected Areas MPAs, localities in the marine environment where environmentally damaging activities, such as certain types of fishing, are restricted.

marine snow Sticky organic material that clumps together and falls like snowflakes down the water column to the ocean floor.

medusa In Coelenterates, the free-swimming body form that resembles an umbrella or bell; it is the main body form of jellyfish.

microbial carbon pump Microbe-dependent activity that causes long-term storage of carbon by transferring unstable forms of organic matter into more durable forms.

microbial loop In marine biology, a pictorial representation of the cycling of organic carbon, nutrients, and energy through microbial communities (viruses, bacteria, and other single-celled organisms) in the marine environment.

mid-ocean ridge An underwater volcanic mountain range formed by the movement of tectonic plates.

neap tides A period characterized by relatively low tidal range that occurs twice each lunar month.

nekton Animals, such as fish and squid, that are strong enough swimmers to be able to move against a water current.

neritic zone The shallow part of the marine environment extending from mean low tide level to a depth of 200 metres.

nitrogen-fixing bacteria Bacteria capable of incorporating atmospheric nitrogen into inorganic compounds that can be used by photosynthesizing organisms.

Ocean Conveyor Belt A system of ocean currents that transports water and heat around the globe.

ocean trenches Long, narrow depressions in the sea floor that form the deepest part of the marine environment.

oceanic zone The area of the ocean lying beyond the continental shelf, typically assumed to begin at water depths below 200 metres.

oxygenic photosynthesis Photosynthetic reactions that involve the production of free oxygen as well as organic carbon.

Particulate Organic Carbon Generally, organic matter that does not pass through a filter of pore size 0.22–0.7 micrometres.

PCR See Polymerase Chain Reaction.

peer-review A process of subjecting an author's scholarly work, research, or ideas to the scrutiny of others who are experts in the same field.

pelagic Pertaining to the open sea; it includes the whole water column.

phagocytosis 'Cell eating'; the process by which cells engulf another cell or a large particle.

phenology The study of cyclic and seasonal phenomena, such as breeding and migration in marine organisms, especially in relation to climate.

photic zone The region of the marine environment close to the sea surface where there is enough light for photosynthesis.

Photosynthetically Active Radiation PAR, the name given to the portion of the solar radiation that can be used for photosynthesis.

Phylogenetic Species Concept The idea that regards a species as a group whose members have a shared and unique evolutionary history, are descended from a common ancestor, and which possess a combination of characteristic and defining traits.

phytoplankton Photosynthetic organisms that drift or float passively in a body of water.

polar Applied in chemistry to a molecule or some other structure that has an uneven distribution of electron density.

Polymerase Chain Reactions PCR, the method used to rapidly replicate specific strands of DNA.

polyp In marine biology, a form of a coelenterate, such as a sea anemone, that has a columnar body with the mouth uppermost, surrounded by tentacles.

porphyria A group of liver disorders in which porphyrins (a class of pigments that form part of haemoglobin and chlorophyll) build up in the body and affect the skin or nervous system.

prevailing wind The most common and persistent wind in a particular region of the world.

protocell A self-organized, spherical structure consisting of a double layer (bilayer) of lipids.

provisioning service A type of ecosystem service that is of direct benefit to human populations and which are commonly traded and have obvious economic value.

pure research A systematic study carried out to increase knowledge and deepen the theoretical understanding of a phenomenon, and which is not intended to be of immediate commercial or practical value.

radula The flexible, coiled ribbon-like structure with chitinous teeth used by grazing molluscs to scrape algal food off rocks, and by predatory snails, such as dogwhelks, to bore holes in shells.

reductionist A person who analyses and describes a complex phenomenon in terms of its simple or fundamental parts.

regulating service An ecosystem service that moderates a natural phenomenon, for example, by reducing the harmful effects of pollutants.

relict organism Applied to an organism that has survived while other related ones have become extinct.

Reynolds number A dimensionless constant that relates to the resistance to movement through a fluid.

salinity A measure of the amount of total concentration of dissolved solids in seawater.

seamount An underwater mound, often of volcanic origin, that rises steeply from the ocean floor.

seaweeds Common name given to any of the red, green, or brown macroalgae (large algae) that grow on the shore and in the shallow seas.

Secchi disc A disc, usually with black and white quarters, used to assess water quality by recording the depth at which the disc is no longer visible from the surface.

Simpson diversity index A measure of species diversity that takes into consideration both the number of species and the number of individuals within each species.

sink In ecology, applied to a population that is a net importer of individuals.

source In ecology, applied to a population that is a net exporter of individuals.

species The basic unit of classification and taxonomic rank of an organism.

species diversity The number of species and their relative abundance in a defined area.

species evenness A measure of biodiversity that focuses on the relative abundance of different species in a sampling area in terms of how close in number each species in a community is.

specific heat capacity The energy required to raise the temperature of one kilogram of a material by one degree Celsius.

spring tide A period with a relatively large tidal range which typically occurs twice a lunar month.

sublittoral zone Area of the marine environment that extends from the lowest point of ordinary tides to the end of the continental shelf.

supporting services An ecosystem service that underpins all other ecosystem services by maintaining the conditions necessary for life.

surface tension A physical property of the interface between a liquid and gas, or between a liquid and solid, that makes the interface behave as if its surface is enclosed in an elastic skin. The surface tension of a seawater–air interface is exceptionally high because water molecules at the interface form hydrogen bonds with other water molecules around and below them, but not with the air molecules above.

taxonomist A biologist who groups organisms into categories, including genus and species.

thermohaline circulation The relatively slow movement of water in the deep ocean produced by density differences of the water that are caused by variations in temperature and salinity

tidal currents Horizontal movements of seawater associated with tides.

topography In the marine environment, the way the natural and physical features of the seafloor are arranged.

Total Economic Valuation A framework for systematically identifying and estimating all the financial benefits of a marine ecosystem service.

transmission electron microscope A type of electron microscope in which an image is derived from electrons that pass through a specimen.

type specimen The specimen on which the description, name, and classification of a new species is based.

upwelling The slow, upward movement of water from the deep sea towards the surface.

virion The entire virus particle, consisting of an outer protein shell and an inner core of nucleic acid, that occurs outside a host cell.

viscosity A measure of a fluid's resistance to flow.

whalefall The sinking of the carcass of a whale from upper water layers down onto the sea floor.

zooplankton Animals and protozoans in the plankton.

zooxanthellae Symbiotic single-celled algae that live in the tissue of some anemones and coral.

SUBJECT INDEX

Note: Tables and figures are indicated by an italic *t* or *f* following the page number.